SpringerBriefs in History of Science and Technology

More information about this series at http://www.springer.com/series/10085

Michael Eckert

The Turbulence Problem

A Persistent Riddle in Historical Perspective

Springer

Michael Eckert
Forschungsinstitut
Deutsches Museum
Munich, Germany

ISSN 2211-4564   ISSN 2211-4572   (electronic)
SpringerBriefs in History of Science and Technology
ISBN 978-3-030-31862-8   ISBN 978-3-030-31863-5   (eBook)
https://doi.org/10.1007/978-3-030-31863-5

This Springer imprint is published by the registered company Springer Nature Switzerland AG
The registered company address is: Gewerbestrasse 11, 6330 Cham, Switzerland

Preface

Turbulence belongs to the realm of fluid dynamics, a discipline founded on the solid pillars of classical mechanics. Its basic equations, the Navier–Stokes equations, have been established in the nineteenth century. Yet it is regarded as the last major unsolved problem of classical physics. The turbulence problem rose to prominence as one of the most persistent challenges of science. The eddies in the turbulent flow of a river or the smoke from a chimney elude a physical understanding from first principles.

In the course of the twentieth century, turbulence became a research field where high expectations met with recurrent frustration. This makes turbulence an ideal subject for the historian of science and technology. On the route towards a history of turbulence, this book is focused on what the actors in this research field perceived as the turbulence problem. At different times and in different social and disciplinary environments, the nature of this problem changed in response to changing research agendas.

When the participants in this quest review their research field, they focus on the progress made for solving the riddles of turbulence. In contrast to participants' reviews, my emphasis is rather on the broader context in which the turbulence problem(s) became enunciated. I am aiming for historical authenticity by quoting as far as possible from contemporary sources (letters, reports, papers). If the original quote was in German, I translated it in English (indicated by "Translation ME" in the footnote). My narrative is descriptive and proceeds in chronological order from around 1900 to the last decade of the twentieth century, so that one or another variant of the turbulence problem will be revisited in subsequent chapters in different circumstances. I do not aim at a comprehensive account but rather at an exemplary exposition of the environments in which problems become items of research agendas. From this perspective, the turbulence problem also provides more general lessons for the history and epistemology of science and technology in the twentieth century.

Munich, Germany
August 2019

Michael Eckert

Contents

Abbreviations

AIP	American Institute of Physics, College Park, MD
APS	American Physical Society
ASWB	Arnold Sommerfeld. Wissenschaftlicher Briefwechsel. Band I: 1892–1918; Band II: 1919–1951. Herausgegeben von Michael Eckert und Karl Märker. München, Berlin, Diepholz: Deutsches Museum und GNT-Verlag, 2000 und 2004
DFD	Division of Fluid Dynamics of the American Physical Society
DFDA	Division of Fluid Dynamics of the American Physical Society, Archives, Lehigh University, Bethlehem, Pennsylvania
DLR	Deutsches Zentrum für Luft- und Raumfahrt
DMA	Deutsches Museum, Archiv, München
GAMM	Gesellschaft für Angewandte Mathematik und Mechanik
GOAR	Historical Archive of the DLR, Göttingen
IAS	Institute of the Aeronautical Sciences, New York
IAU	International Astronomical Union
IUGG	International Union of Geodesy and Geophysics
IUTAM	International Union of Theoretical and Applied Mechanics
NACA	National Advisory Committee for Aeronautics, Washington, D.C.
NPL	National Physical Laboratory, Teddington
RANH	Rijksarchief in Noord-Holland, Haarlem
SUB	Niedersächsische Staats- und Universitätsbibliothek, Göttingen
TKC	Theodore von Kármán Collection, California Institute of Technology, Pasadena
ZAMM	Zeitschrift für Angewandte Mathematik und Mechanik
ZWB	Zentrale für wissenschaftliches Berichtswesen der Luftfahrtforschung des Generalluftzeugmeisters

List of Figures

Chapter 1
Hydrodynamics Versus Hydraulics

Abstract In the beginning of the 20th century the study of fluid motion fell into one of two classes: either the observed flow could be described in terms of hydrodynamics, or it eluded theory and belonged to the realm of hydraulic engineering, like pipe- or channel flow. The discrepant results of hydrodynamics versus hydraulics illustrated the gulf between theory and practice—and turbulence was regarded as the culprit. The rise of aeronautics added further challenges. Wind tunnel investigations hinted at turbulence effects that eluded theoretical analysis.

Research on turbulence has no clear-cut beginning. The history of quantum mechanics, by comparison, may be confined to the time span between Max Planck's formula for black-body radiation in 1900 and the mid 1920s when Werner Heisenberg, Erwin Schrödinger and others established matrix- and wave mechanics. The history of turbulence has no such landmarks.

1.1 When and How Turbulence Became a Problem

Turbulent flow must have been observed since antiquity. It was surely perceived as strikingly different from the smooth flow regime later called laminar. The earliest observations of turbulence that left a trace in historic records are due to Leonardo da Vinci whose sketches of eddying water flow are frequently used in reviews on turbulence.[1] With the rise of "rational fluid mechanics" (Truesdell 1954) in the 18th century flow phenomena became subject of mathematical analysis. Johann and Daniel Bernoulli, Leonhard Euler, Jean-Baptiste le Rond d'Alembert and other pioneers of ideal flow theory must have been aware that there was a fundamental mismatch between theory and experiment which nearly always concerned eddying flow. "Euler's equation" for ideal flow was extended in the first half of the 19th century to

---

[1]See, for example, Tennekes and Lumley (1972), Frisch (1995), Ecke (2005) and websites such as http://sersol.weebly.com/workshop/leonardo-da-vinci-and-physics-of-fluid (24 October 2019).

© The Author(s), under exclusive license to Springer Nature Switzerland AG 2019
M. Eckert, *The Turbulence Problem*,
SpringerBriefs in History of Science and Technology,
https://doi.org/10.1007/978-3-030-31863-5_1

the "Navier-Stokes equation" by taking into account fluid friction—but that did not immediately foreground turbulence as a particular problem of hydrodynamic theory.

But hydraulic engineers concerned with the design of channels and water pipes had made turbulent flow already earlier a subject of experimental and theoretical studies, by . The history of turbulent flow formulae for the discharge of water in pipes and channels starts in the 18th century with Antoine Chézy, Albert Brahms and others. Thus a tradition of hydraulic flow studies was launched that resulted in well-established laws of turbulent flow—long before theory was up to derive them from the Navier-Stokes equation.[2] The first researcher who discerned the transition to turbulence as the culprit for discrepant flow laws was Gotthilf Hagen, an hydraulic engineer famous for his investigations published in 1839 of laminar pipe flow ("Hagen-Poiseuille law"). Fifteen years later Hagen reported about observations of "two kinds of motions"—similar to those for which Osborne Reynolds entered the hall of fame in turbulence more than three decades later (see below). Hagen visualized the transition to turbulence with sawdust added to a flow of water in a glass tube. "I observed that for small pressures the sawdust propagated only in the direction of the tube, whereas for strong pressures it shot from one side to another and often assumed a vortical motion."[3]

From a theorist's vantage point an early expression of what would later be called the "turbulence problem" is due to Adhémar Barré de Saint-Venant. "The problem to establish in each case the differential equations for the motion and to integrate them has still its often great difficulty," he concluded in 1872 a study "On the hydrodynamics of streams of water" in which he reviewed recent efforts, among others, of his pupil Joseph Boussinesq. In contrast to the older theories of Navier and others Boussinesq's approach was based on recent experiments by hydraulic engineers. "From this respect," Saint-Venant concluded, "the problem represents no longer this hopeless enigma which the distinguished minds have attacked in vain."[4]

When Osborne Reynolds published in 1883 his momentous study in which he established the concept of what later was named Reynolds number, turbulence had long been investigated and perceived as a riddle—even though in vague terms only. Reynolds introduced his treatise with the remark that the results of his investigation "have both a practical and a philosophical aspect." The former concerned the law for the resistance in pipe flow, the latter the transition from laminar to turbulent flow, or, to speak with Reynolds, "a definite verification of two principles, which are—*that the general character of the motion of fluids in contact with solid surfaces depends on the relation between a physical constant of the fluid and the product of the linear dimensions of the space occupied by the fluid and the velocity.*" He established the criterion which discriminates the "direct" from the "sinuous" flow, to use his own terms for the distinction between laminar and turbulent flow—but

---

[2]On the history of hydraulics see Rouse and Ince (1957); for a review of turbulent pipe and channel flow laws see Hager (1994).

[3]Hagen (1854, p. 81), quoted from Darrigol (2002, pp. 241–242).

[4]Saint-Venant's and Boussinesq's approach on "tumultuous water" is reviewed in Darrigol (2002, pp. 220–231).

without the rhethoric of great enigma or riddle (Reynolds 1883). Nor did George Gabriel Stokes and Lord Rayleigh, who acted as referees for the Royal Society, praise the paper as a major contribution in the quest for a theory of turbulence. Rayleigh's verdict comprised 70 words and merely acknowledged "that the results are important, and that the paper should be published in the Phil. Trans." With regard to the theoretical implication of Reynolds' experimental findings he concluded that "the Author refers to theoretical investigation whose nature is not sufficiently indicated." Stokes maintained that "there is absolutely nothing to prove that he has discovered the necessity of an additional constant to define a fluid." (Launder and Jackson 2011, pp. 20–21).

A year later Reynolds presented a popular lecture at the Royal Institution "On the two manners of motion of water". Without explicit reference to turbulence he asserted "that in spite of the most strenuous efforts of the ablest mathematicians, the theory of fluid motion fits very ill with the actual behaviour of fluids; and this for unexplained reasons" (Reynolds 1884, p. 42). Ten years later he made himself a strenuous effort to explain the apparent impotence of the fundamental equations by adding expressions that were later called "Reynolds stresses" (Reynolds 1895). But once more he received only a lukewarm response. Stokes' referee report was neither positive nor negative; he more or less admitted that he did not understand the argument—which is little surprising in view of Reynolds' sophisticated distinctions between "mean-mean" and "relative-mean" motions. Sir Horace Lamb, the other referee, recommended the publication of Reynolds' paper "although much of it is obscure and there are some fundamental points which are not clearly established."[5] Reynolds himself does not seem to have been fully convinced that he had achieved a major accomplishment with regard to the riddle of turbulence.

1.2   Felix Klein's Efforts to Bridge the Gulf Between Hydraulics and Hydrodynamics

The studies of Saint-Venant, Boussinesq and Reynolds clearly show that turbulence was recognized at the end of the 19th century as the culprit for the gulf between hydrodynamics and hydraulics. But what later would be named as "turbulence problem" was only implicitly addressed as subject for particular investigation. Even the term "turbulence" made its appearance only gradually (Schmitt 2017).

At the turn from the nineteenth to the twentieth century, the gulf between theory and practice in fluid mechanics reached its climax. Wilhelm Wien, then a theoretical physicist and renowned expert on hydrodynamics, introduced a textbook on this subject with the observation that practical applications were to such an extent beyond the reach of theory that engineers adopted their own procedure to deal with hydrodynamical problems, "which is usually called hydraulics." But in Wien's view hydraulics

[5]Quoted in Launder and Jackson(2011, p. 25); for a discussion of Reynolds' paper see Darrigol (2002, pp. 259–262).

lacked "so much of a strict method, in its foundations as well as in its conclusions, that most of its results do not deserve a higher value than that of empirical formulae with a very limited range of validity."[6]

It is not accidental that the gulf between theory and practice was most pronounced in Germany, where hydraulics was taught at technical colleges (Technische Hochschulen) and hydrodynamics at universities. The renowned *Encyklopädie der mathematischen Wissenschaften*, edited by the mathematician Felix Klein, assigned the articles on hydrodynamics and hydraulics to different authors (Love 1901a, b; Forchheimer 1905). Klein was eager to bring appied sciences to universities—much to the chagrin of engineering professors who struggled for the emancipation of technical colleges as equal to the universities. They considered Klein's flirtation with applied sciences as an intrusion in their own territory (Manegold 1970). To cut a long story short: applied sciences, like hydraulics, were a minefield in the German academia. Klein's lectures and seminars reflect this atmosphere. He presented hydraulics as a discipline in need for scientific underpinning. By the same token he criticised the contemporary tendency among mathematicians to shun applications in favour of pure theory. "Two tendencies are nowadays prevailing," Klein introduced a lecture on hydrodynamics in autumn 1899, "the tendency towards practical application on one side, and on the other the striving to develop the pure theory as far as possible."[7]

In this context Klein presented turbulence as a particular challenge. He derived "the so-called Poiseuille laws" for capillary tubes from the Navier-Stokes equations and contrasted them with empirical formulae for pipe flow in wider tubes. "What is responsible for this discrepancy? We find the answer in a fundamental paper by Osborne Reynolds in the Phil. Trans. of 1883," Klein referred to this study in which Reynolds had shown that beyond a critical velocity "the entire fluid was filled with numerous irregular vortices. These motions are called 'turbulent' motions." Klein argued that beyond the critical velocity the laminar flow becomes unstable[8]:

> Why the instability occurs is unknown, and one has to confine oneself to devise the laws of turbulent flow oneself by a combination of observation and thought. Mathematicians and engineers are extraordinarily unable to communicate with another in the field of fluid motions and related questions because the instability of flow has always been neglected by the mathematicians, while the engineers mostly had to do with instability without knowing it clearly. In the best case frictional terms have been added to the equations of ideal hydrodynamics, but these completed equations are in most technical problems to no avail—such as in the flow of water through a pipe or the outflow from a faucet, for example—just because the flow in water pipes is to the highest degree turbulent. The fact that turbulent motions occur forms the bridge between hydrodynamics and hydraulics.

Klein's role model was Boussinesq whose recent "Théorie de l'écoulement tourbillonnant et tumultueux des liquides dans les lits rectilignes a grande section" (Boussinesq 1897) attempted "to bring in line the phenomenon of turbulence with

---

[6]Wien (1900, p. III), Translation ME.

[7]Klein's lecture on hydrodynamics, winter semester 1899/1900, elaborated by Karl Wieghardt, p. 1, first lecture, 24 October 1899. Translation ME. On Klein's biography see Tobies (2019).

[8]Ibid., pp. 313–327. Translation ME.

the hydrodynamic equations." Klein dedicated his lecture on 18 January 1900 to a review of Boussinesq's approach which he regarded as "a good beginning."[9]

Three years later turbulence was again on Klein's agenda. Together with Karl Schwarzschild, then professor of astronomy at Göttingen University and the director of the Göttingen observatory, he organized in the winter semester 1903/04 a seminar on "Selected chapters of hydrodynamics." The motivation came from "the hydraulic review" which Philipp Forchheimer, an Austrian engineer, was preparing for the volume on mechanics of the *Encyklopädie der mathematischen Wissenschaften* (Forchheimer 1905). But Klein's intention was more ambitious. He strove for an "all-over understanding of individual parts of mechanics by taking into particular consideration engineering," as he noted in a draft about his aims. He regarded it a "true need of our times to bridge separate developments."[10] Among these issues turbulence deserved top rank in Klein's view. "To consider the circumstances of turbulence from all sides has to be a major chapter," he introduced the seminar.[11]

The seminar contributions on turbulence became subsequently the subject of a common paper by Schwarzschild and the mathematicians Hans Hahn and Gustav Herglotz. They shared Klein's admiration of Boussinesq (Reynolds is only mentioned in passing with reference to his investigation from 1883) and presented an "inductive" approach to the eddy viscosity ε which Boussinesq had introduced in an "apriori-deductive" manner, "for which the field does not yet seem ripe." Boussinesq's derivation of differential equations for mean velocities was based on the assumption that the "apparent friction of turbulence" was due to a transformation of kinetic energy into heat as a result of the frequent changes of flow velocities. With reference to a study of Hendrik Antoon Lorentz (1897) Schwarzschild and his co-authors refuted this view. As Lorentz had shown "the augmented apparent friction is due to transport and exchange of momentum that takes place between the individual sites in a section of the stream in turbulent motion." Nevertheless they regarded Boussinesq's eddy viscosity ε—defined as a variable coefficient of friction proportional to the energy of the turbulent motion—a salient feature of a future theory of turbulence. Their focus on this quantity was so strong that they came close to identify it with turbulence itself: "... and we allow ourselves to designate ε as 'turbulence' for short."[12] Yet Schwarzschild, Hahn and Herglotz did not expound the difficulty to express ε as a function of actually measured quantities of turbulent pipe and channel flows as the turbulence problem—they postponed it for further investigations with suspended particles in eddying flows from which they expected more direct insight into the nature of turbulence.

---

[9]Ibid., pp. 327–334. Translation ME.

[10]Klein's notes for the "Seminar 1903/04", fol. 16 ("Eröffnungsvortrag meinerseits, 28. Oktober 1903"). SUB, Handschriften, Cod. Ms. F. Klein 19E. Translation ME.

[11]Seminar on "Ausgewählte Kapitel der Hydrodynamik, WS 1903/04". The protocols are available online at http://www.uni-math.gwdg.de/aufzeichnungen/klein-scans/klein/V20-1903-1904/V20-1903-1904.html. (24 October 2019). Translation ME.

[12]Hahn et al. (1904, pp. 412–415). Translation ME. For a discussion from a modern perspective see Kluwick (1996).

Fig. 1.1 Ludwig Prandtl, here in 1904 in front of a water tank built for studying the formation of vortices, made Göttingen a center for research on turbulence

Four years later Klein organized again a seminar on hydrodynamics. "General intent: Reconciliation of theory and practice," he introduced the participants of this seminar once more to his aims.[13] This time Klein expressed his intent to collaborate with applied sciences by involving as co-organizers Emil Wiechert, the professor of geophysics, Ludwig Prandtl (Fig. 1.1) and Carl Runge. Prandtl an Runge had been called to Göttingen University in 1904 and directed the institutes for applied mechanics and mathematics, respectively. Their advanced students bore the brunt of the seminar presentations.[14]

Prandtl, in particular, was keenly interested in turbulence.[15] It is not accidental, therefore, that turbulence was again discerned as a major theme. It was reviewed at two sessions, on 29 January and 5 Februar 1908, by Heinrich Blasius, Prandtl's first doctoral student, and it reveals an intimate familiarity with the pertinent literature on hydraulics as well as hydrodynamics. Although Blasius, like his predecessors, did not explicitly expound "the turbulence problem" as a particular riddle, he demonstrated in great detail where hydraulics and hydrodynamics diverged. The mere disposition

---

[13]Eckert (2019b). Translation ME. Klein's notes on "Hydrodynamisches Seminar 1907/08," fol. 2 ("I. Sem. 30. Oktober 07.") SUB, Handschriften, Cod. Ms. F. Klein 20F.

[14]The protocols are available online at http://www.uni-math.gwdg.de/aufzeichnungen/klein-scans/klein/V27-1907-1909/V27-1907-1909.html. (24 October 2019). For transcriptions and comments see Eckert (2019b).

[15]For Prandtl's biography see Eckert (2017, 2019a).

how he structured the theme of "turbulent flows" reflects an acute awareness of the involved problems. The first topics largely concerned empirical observations such as the different velocity profiles of laminar and turbulent pipe flow. From these observations Blasius concluded that hydrodynamics had to answer the following questions: "I. Explanation of instability, II. Representation of turbulent motion." The latter would have to distinguish between the different answers of "hydraulics" and the "rational hydrodynamical approaches."[16]

1.3 The Turbulence Problem in the Early Sommerfeld School

Blasius, like four years earlier Schwarzschild, Herglotz and Hahn, ignored the theoretical effort with which Reynolds had attempted to account for turbulent fluctuations in the averaged Navier-Stokes equations (Reynolds 1895). Nor did Reynolds's theory enter Wilhelm Wien's textbook on hydrodynamics published in 1900. "Reynolds not only performed superb experiments which you cite, but also a magnificent theory," Arnold Sommerfeld criticized Wien for this neglect. "I regard this work by Reynolds the greatest achievement in hydrodynamics since the establishment of the differential equations... I have studied these things for my lecture and hope that I can pursue them further."[17]

Arnold Sommerfeld had earned his spurs as assistent of Felix Klein in the mid 1890s. In 1900 he was appointed as professor of mechanics to the Technische Hochschule Aachen. He chose hydraulics as a research subject in order to counter the distrust of his colleages from the engineering departments who regarded mathematicians from universities—and a disciple of Klein in particular—with suspicion. The opportunity to demonstrate his readiness to deal with practical matters came in September 1900 when the Gesellschaft deutscher Naturforscher und Ärzte convened their annual meeting in Aachen. Under the title "Recent investigations on hydraulics" Sommerfeld announced his intent to study engineering applications such as the lubrication of machines for which the "physical theory of fluid friction" seemed appropriate. But he was quite explicit that the "physical theory", i.e. hydrodynamics, was often in conflict with the "technical theory", i.e. hydraulics. He exposed the discrepant formulae for laminar and turbulent pipe flow as a major challenge for a reconciliation of both approaches[18]:

> Physical theory predicts a frictional resistance proportional to the velocity and inversely proportional to the square of the diameter, according to the technical theory it is proportional to the square of the velocity and inversely proportional to the diameter. The physical theory agrees splendidly in capillary tubes; but if one calculates the frictional losses for a water pipeline one finds in certain circumstances values which are wrong by a factor of 100.

---

[16]Eckert (2019b, p. 131). Translation ME.

[17]Sommerfeld to Wien, 30 June 1900. DMA, NL 56, 010. Translation ME.

[18]Sommerfeld (1900). Translation ME.

The first theoretical attempt to resolve the discrepancy from first principles was Reynolds's paper from the year 1895. "In this important work the theory of viscous fluids is subjected to a comprehensive revision. How necessary this is becomes evident from observations of fluid motions in pipes," Sommerfeld had briefly characterized the scope and motivation of this effort in the *Jahrbuch über die Fortschritte der Mathematik*.[19] Shortly after the Aachen meeting, in October 1900, Sommerfeld confessed in a letter to Lorentz, that he had failed in his own attempts "aimed at a determination of possible motions above the critical velocity."[20] In this venture Sommerfeld also scrutinized Lorentz's own effort to extend Reynolds's work by a more refined consideration of the turbulent velocity fluctuations and noted an error which Lorentz duly acknowledged in a later revision (Lorentz 1907, p. 63). Sommerfeld hoped to contribute with a theory about the onset of turbulence to the *Festschrift* for celebrating the 25th anniversary of Lorentz's doctoral dissertation, but, as he wrote two months later, he "miserably shipwrecked" in this effort.[21]

Soon thereafter Sommerfeld enounced once more his concern about the gap between hydrodynamics and hydraulics. It seems "that theoretical hydrodynamics would have to declare itself bankrupt in view of the practical problems of hydraulics," he sighed at another annual meeting of the Gesellschaft deutscher Naturforscher und Ärzte. "Still there is no precise theoretical method to determine the critical velocity and the pressure gradient beyond the critical velocity."[22] The focus on the energy of the velocity fluctuations from which Reynolds and Lorentz hoped to derive a critical limit for the transition to turbulence did not yield the desired result. Instead of this "energy method," as it was later called, Sommerfeld resorted to another approach with which he had become familiar in the theory of buckling. "The buckling of the plate had an interesting sequel," he wrote to Carl Runge. "I noticed that a similar calculation also leads to a determination of the critical velocity in hydrodynamics. For the time being, however, I am left with a rather horrible transcendental equation that awaits further discussion."[23] But his hopes were frustrated. "Unfortunately I could not make progress with the problem to determine the critical velocity in hydrodynamics," he confessed in a letter to Lorentz.[24]

By this time Sommerfeld had been called to Munich University on the chair for theoretical physics. It had been created for Ludwig Boltzmann in 1890 but left vacant since Boltzmann had left Munich in 1894. The prestigious legacy must have kindled Sommerfeld's determination not to shy away from a problem that even Lorentz— then the uncrowned master of theoretical physics—could not solve. "I hope to bring my old efforts on 'turbulence' to an end," he announced in a letter to Lorentz in March

---

[19] JFM 26.0872.02, available online at www.emis.de. (24 October 2019). Translation ME.

[20] Sommerfeld to Lorentz, 8 October 1900. RANH, Lorentz, inv.nr. 74. Also in ASWB I, pp. 180–182. Translation ME.

[21] Sommerfeld to Lorentz, 10 December 1900. RANH, Lorentz, inv.nr. 74. Translation ME.

[22] Sommerfeld (1903, p. 212). Translation ME.

[23] Sommerfeld to Runge, 9 June 1906. DMA, HS 1976-31. Translation ME.

[24] Sommerfeld to Lorentz, 12 December 1906. RANH, Lorentz, inv.nr. 74. Also in ASWB I, pp. 257–258. Translation ME.

1908 a forthcoming paper.[25] Lorentz praised Sommerfeld for his productivity.[26] But Sommerfeld could not proceed considerably beyond the "horrible transcendental equation" at which he had arrived already earlier. He failed once more to derive a critical limit beyond which laminar flow becomes turbulent. "I have tortured myself continually with work on turbulence and spent almost all of my time with it, but I could not accomplish it," he wrote to Wilhelm Wien three months later. "A short preliminary note appears in the Rome proceedings."[27]

The latter remark referred to a paper which Sommerfeld had submitted to the Fourth International Congress of Mathematicians, held in April 1908 in Rome. Sommerfeld did not attend the Congress himself, nevertheless his paper was included in the proceedings (Sommerfeld 1909). Together with a study by William McFadden Orr, who had published in 1907 an extensive treatise on the stability of laminar flows (Orr 1907), these papers mark the beginning of the so called Orr-Sommerfeld approach. It is beyond the scope of this chapter to review this approach,[28] but it seems appropriate to quote a passage from Sommerfeld's note that explicitely expounds "the problem of turbulence"[29]:

> The condition of instability appears in the form of a transcendental equation. The present communication only leads to the establishment of this equation; I have not yet accomplished its complete discussion which I regard as the true content of the problem of turbulence.

Henceforth Sommerfeld identified the problem of turbulence with the determination of the critical limit beyond which laminar flow becomes unstable—and as such it emerged as a challenge for some of his disciples. Ludwig Hopf, one of Sommerfeld's first doctoral students, titled the first paragraph of his thesis on "Hydrodynamic investigations" with the headline "The turbulence problem." The "consistent treatment of the problem according to the method of small oscillations by Sommerfeld is not yet accomplished," Hopf introduced his study.[30] But he did not attempt to proceed with the problem beyond the stage where Sommerfeld had left it—that was reserved for a later investigation. Apparently Sommerfeld regarded the problem too much of a challenge for a doctoral dissertation. Instead he assigned to Hopf the task to determine the transition to turbulence in open channel flow experimentally. "There is under certain circumstances (small depth, slow flow velocity) a stable laminar motion of the kind of Poiseuille's law, which becomes unstable beyond a critical limit and gives rise to another form of motion," Sommerfeld summarized the goal of

---

[25] Sommerfeld to Lorentz, 18 March 1908. RANH, Lorentz, inv.nr. 74. Also in ASWB I, pp. 331–332. Translation ME.

[26] Lorentz to Sommerfeld, 27 March 1908. DMA, HS 1977-28/A,208. Also in ASWB I, pp. 333–334.

[27] Sommerfeld to W. Wien, 20 June 1908. DMA, NL 56, 010. Also in ASWB I, pp. 341–343. Translation ME.

[28] Sommerfeld's approach is reviewed in more detail in Eckert (2010). For a comprehensive study see Drazin and Reid (2003) and (Yaglom 2012).

[29] Sommerfeld (1909, p. 117). Translation ME.

[30] Hopf (1910, p. 7). Translation ME.

Hopf's experiments.[31] Hopf used a straight 5.2 cm wide brazen water channel and varied the flow speed by increasing the channel's inclination. He also modified the viscosity by adding sugar. "I still remember how I brought him a whole sack-full of sugar so he could make a sugar solution," Sommerfeld's assistant recalled many years later Hopf's experiment on turbulence.[32] Although Hopf determined critical limits, his results were not really conclusive because capillary waves interfered with the transition to turbulence (Hopf 1910, p. 25).

The true "turbulence problem" as enunciated by Sommerfeld and Hopf amounted to the solution of the transcendental equation derived in Sommerfeld's Rome paper. Although Sommerfeld had chosen the simplest case of a plane flow between two parallel plates moving at constant speed in opposite direction (plane Couette flow), the transcendental equation from which the critical Reynolds number would be determined was a monster. Hopf tackled with this equation in the next step of his academic career: "The development of small oscillations in viscous flows" became the subject of his habilitation thesis. He must have been disappointed about the result of his laborious mathematical effort because it meant that plane Couette flow would not experience a transition to turbulence. Any small disturbances superimposed to the laminar flow would vanish exponentially and not induce turbulence (Hopf 1914, p. 59). Sommerfeld had apparently obtained the same result in some unpublished work because Prandtl responded to him in April 1911: "your result on turbulence has interested me very much. So the dreaded stability indeed has occurred!"[33]

Despite—or because of—this result, the "turbulence problem" became a challenge for other theorists from Sommerfeld's circle. In May 1912 Sommerfeld presented to the Bavarian Academy of Science a treatise "On the expansion of an arbitrary function by eigenfunctions of the turbulence problem" submitted by the mathematician Otto Haupt, who spent a one-year sojourn in Munich. The result confirmed Sommerfeld's and Hopf's earlier verdict about the "dreaded stability". "A hydrodynamical explanation of the turbulence phenomena in terms of the method of small oscillations, therefore, appears impossible," Sommerfeld concluded.[34] But in the following year he presented to the Academy another paper which seemed to bypass the stability deadlock. The author of this paper was Fritz Noether, like Hopf an early student of Sommerfeld with a strong tendency towards mathematics. Instead of the linear velocity profile of plane Couette flow, Noether chose a cubical parabola as initial velocity profile—and obtained a critical limit of stability. With different initial states of flow, Noether argued, "the theoretical treatment of the turbulence problem" must also be different.[35] But at a subsequent session Sommerfeld presented to the Academy a study "On the turbulence problem" by Otto Blumenthal, a friend and colleague who

---

[31] Sommerfeld's report to the philosophical faculty of Munich University, 5 July 1909. Munich University Archive (OC I 35 p). Translation ME.

[32] Peter Debye, interviewed by T. S. Kuhn und G. Uhlenbeck, 3 May 1962. AIP. Available online at http://www.aip.org/history/ohilist/4568_1.html. (24 October 2019).

[33] Prandtl to Sommerfeld, 5 April 1911. DMA, NL 89, 012. Translation ME.

[34] Haupt (1912, pp. 10*–11*). Translation ME.

[35] Noether (1913, p. 328). Translation ME.

had studied with Sommerfeld in Göttingen, which contradicted Noether's results. "Thus there is still no case known in which a laminar flow can be transformed into a turbulent flow."[36]

1.4 Hydraulics and Turbulence

The quest for a "hydrodynamical explanation of the turbulence phenomena in terms of the method of small oscillations"—to cite Sommerfeld again—could nourish the impression that mathematicians and theoretical physicists isolated the problem from its hydraulic context for the purposes of their own specialty. Few hydraulic engineers would have ventured to follow Noether's advice to make use of the theory of nonlinear integral equations (Noether 1913, p. 329). Yet even mathematicians like Noether and Blumenthal kept close contact with colleagues from engineering departments. Noether was assistant in the institute for mechanics at the Technische Hochschule Karlsruhe. Blumenthal, then professor of mathematics at the Technische Hochschule Aachen, acknowledged that he received the suggestion to scrutinize Noether's theory from Theodore von Kármán who had just been appointed as professor for mechanics and aerodynamics in the neighboring engineering department. Georg Hamel, another mathematician who paid tribute to the turbulence problem (Hamel 1911), was professor of mechanics at the Technische Hochschule Brünn. His assistant, Richard von Mises, published in 1912 a study on "Small oscillations and turbulence" (von Mises 1912) that arrived by a different method at the same result as Sommerfeld and Hopf.

Richard von Mises exemplifies the spirit in which theorists with a tendency for practical applications approached their subjects. His habilitation at the Technische Hochschule Brünn was dedicated to the theory of water wheels, a treatise which already displayed Mises's eagerness to apply mathematics to hydraulic engineering (von Mises 1908a). On the book cover the author added "engineer" to his name. In 1908, at the annual meeting of German mathematicians, Mises presented as a challenge for future work "problems of technical hydromechanics." Here he defined "the so-called turbulence problem" as "the task to find the integral of Stokes's equations that corresponds to the real pulsating motion in cylindrical flow," in other words: to derive from the Navier-Stokes equations solutions that correspond to turbulent pipe flow. Attempts to solve this and other problems may be regarded as preliminary, Mises admitted, but "one must seize occasionally the imperfect if one does not simply wish to ignore the needs of practice."[37]

Despite the often bemoaned gulf between hydraulics and hydrodynamics, some engineers and physicists in hydraulic laboratories responded to the challenge of turbulence by their own means. From a practical perspective, the resistance of turbulent pipe flow was of paramount importance. In 1901, a research effort was launched at the Hydraulics Laboratory of Cornell University in order to measure the "head loss"

---

[36]Blumenthal (1913, p. 564). Translation ME.

[37]Mises (1908b, pp. 323 and 325). Translation ME.

at high precision for a variety of pipe diameters and flow speeds. Although Reynolds
had introduced dimensional considerations twenty years earlier, head loss formulae
were not yet expressed in a standardized way so that different flow measurements
could not easily be compared with another. The name "Reynolds number" was introduced only in 1908 with Sommerfeld's Rome paper (Rott 1990). The measurements
at the Cornell Hydraulics Laboratory, performed by Augustus V. Saph and Ernst
W. Schoder on a variety of smooth brass pipes of different diameters with water at
different flow speed, yielded as a general head loss formula

$$\frac{h}{l} \propto \frac{u^{1.75}}{d^{1.25}},$$

where the head loss h referred to a straight pipe of length l and diameter d and the
flow velocity u. The graphical representation of the data also showed a systematic
dependence on the temperature (measured at 40, 55 and 70° Fahrenheit) (Saph and
Schoder 1903; Steen and Brutsaert 2017).

The data from the Cornell Hydraulics Laboratory became the major source for the
first modern representation of turbulent pipe flow resistance in terms of a dimension-
less friction coefficient λ expressed as a function of the Reynolds number $\frac{ud}{\nu}$. The
best fit to the data from the Cornell Laboratory and to new measurements performed
in the Preussische Versuchsanstalt für Wasserbau und Schiffbau in Berlin yielded

$$h = \lambda \cdot \frac{l}{d} \cdot \frac{u^2}{2g}, \quad \text{with} \quad \lambda = 0.316 \cdot (\frac{ud}{\nu})^{-1/4}.$$

The author of this formula was Heinrich Blasius. He published these results in
1911 in a short version addressed to an audience of physicists (Blasius 1911), and
two years later exhaustively in an engineering journal (Blasius 1913). After fin-
ishing his study with Prandtl in Göttingen, Blasius had moved to the Preussische
Versuchsanstalt für Wasserbau und Schiffbau before he continued his career as
professor of mathematics at an engineering college in Hamburg (Hager 2003).

At the Preussische Versuchsanstalt für Wasserbau und Schiffbau pipe flow measurements and other hydraulic applications became the major concern of Blasius's
research. But his motivation to express the head loss formula in terms of the Reynolds
number was rooted in Göttingen where similarity considerations became a matter of
concern in early wind tunnel investigations at Prandtl's model testing facility. With
the same viscosity of air in the wind tunnel and in real flight, the requirement of
the same Reynolds number amounted to the equality of the product of velocity and
model size. Thus early measurements of friction coefficients of plates and wires were
displayed as a function of this product (Föppl 1911, p. 84). Blasius declared the prin-
ciple of model testing also as fundamental for pipe flow experiments. Reynolds's
law demanded that "the quantity $\frac{ud}{\nu}$ must have the same value for the model and
in reality in order to ascertain similarity".[38] Despite its wide range of applicability
"this 30 years ago established law has hardly made inroads in practice," he criticized

---

[38]Blasius (1912, p. 36). Translation ME.

this neglect in hydraulic engineering. "It would simplify considerably the business of interpolations and make it more secure if one would draw λ from the very beginning as a function of the single variable $\frac{ud}{v}$ and make the interpolations accordingly, instead of dealing, as it has been done up to now, with the dependency on two variables, diameter and velocity, and in addition on the temperature".[39]

Previously the transition to turbulence had been displayed in a host of diagrams which displayed different critical velocities according to the respective pipe diameters and viscosities (varied by changing the temperature). Now, with the coefficient of friction displayed as a function of the Reynolds number $R = \frac{ud}{v}$ like in Fig. 1.2, the data from different measurements collapsed so that the transition to turbulence was revealed as a universal feature between two flow regimes, each with a characteristic dependency on the Reynolds number R: for laminar flow it was proportional to R^{-1}, and for turbulent flow it was proportional to $R^{-1/4}$. "The similarity law shows its importance just because it allows a statement on the turbulent flow," Blasius addressed the turbulence problem from this vantage point.[40]

Pipe flow data as presented by Blasius threw oil to the fire of turbulence theory. "It is not yet known whether the phenomena of turbulent flow in pipes may be developed from the Navier-Stokes differential equations," Noether introduced in September 1913 a talk "On the theory of turbulence" at the annual Naturforscher meeting in Vienna. "As is well known, the laminar flow compatible with these equations is only observed in a narrow range of velocities, while for large velocities occurs the theoretically still unexplained 'turbulent' mode of flow."[41] Noether chose "Stokes's friction formula" for the resistance of a sphere moving at constant speed through a fluid in order to investigate its range of validity (Noether 1914b). Stokes's law could be derived from the Navier-Stokes equations and thus represented another case (next to the law of Hagen-Poiseuille for laminar pipe flow) where hydrodynamic theory accounted for fluid friction up to a critical limit. In this case, however, the range of validity was confined to extremely small velocities and diameters. Lord Rayleigh had concluded already in 1893 "by considerations of dimensions" that the quantity $\frac{Va}{v}$, i.e. the Reynolds number expressed by the diameter a and the velocity V of the sphere, had to be rather small. In the case of water a sphere with a diameter of one millimeter should not move faster than 0.01 cm per second, Rayleigh concluded (Rayleigh 1893, p. 365). Noether carried Rayleigh's argument further and developed an approximation to "Stokes's motion" in terms of the Reynolds number. But he made no attempt to derive a critical limit beyond which Stokes's motion would experience a transition to turbulence.

---

[39]Blasius (1911, p. 1176). Translation ME.

[40]Blasius (1913, p. 22). Translation ME.

[41]Noether (1914a, p. 138). Translation ME.

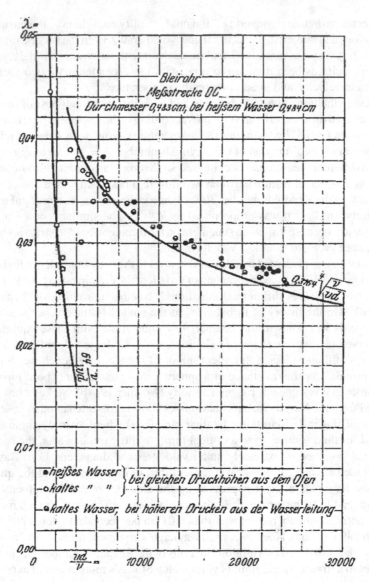

Fig. 1.2 Blasius's diagram for the friction coefficient of laminar and turbulent pipe flow as a function of the Reynolds number (Blasius 1913, Abb. 13)

1.5 Turbulence in the Wake of Spheres and Struts

By the same time the flow around spheres made inroads for the study of turbulence as a result of wind tunnel experiments. Measurements of the air resistance of spheres in Gustave Eiffel's aerodynamic laboratory in Paris suggested that the coefficient of friction suddenly decreased at higher air speeds! Prandtl explained the phenomenon in terms of his boundary layer concept: When the initially laminar boundary layer around the sphere becomes turbulent, it entrains air from the wake behind the sphere so that the boundary layer stays attached to the surface of the sphere longer than in the laminar case. Hence the onset of turbulence in the boundary layer reduces the wake behind the sphere and thus also its drag. The boundary layer concept, which had been applied so far to laminar flow only, thus became extended to turbulent flow. In 1914 Prandtl and his assistant, Carl Wieselsberger, demonstrated this effect by inducing turbulence in the boundary layer with a thin "trip wire" around the sphere. The reduced extension of the wake was made visible with smoke. Thus it was obvious that the trip wire reduced also the air resistance (Rotta 1990, pp. 83–85; Bodenschatz and Eckert 2011, pp. 45–46; Eckert 2017, pp. 88–91).

The Göttingen experiments were not only aimed at a qualitative demonstration. From the very beginning the model tests went hand in hand with similarity considerations. When Prandtl and Wieselsberger checked Eiffel's observation they were very careful to display both their own and Eiffel's measurements of the coefficient of friction as a function of the Reynolds number (Fig. 1.3). Thus it turned out that the effect occurred in Eiffel's wind tunnel at a lower Reynolds number than in Prandtl's wind tunnel. "Because as causes for the drop we suggested the transition to turbulence in the boundary layer, an explanation for this difference was easily found," Prandtl argued. "Like in pipe flow, the turbulence had to occur at lower velocities if there were already vortices in the arriving air-stream." This made spheres indicators of the degree of turbulence in the air-stream of different wind tunnels. "As a measure for comparison" Prandtl suggested "for example, the Reynolds number at which the coefficient of a technically smooth sphere of copper (polished with finely granulated emery cloth) becomes 0.18".[42]

The turbulence effect observed with spheres concerned also other bodies in an air-stream. Prandtl closed his paper with the remark that "one should pay attention to the changeover of the law of resistance for cylindrical poles of circular or ellipsoid cross section".[43] Contemporary airplanes were affected by this phenomenon in a peculiar way because the wings of a biplane—the most common type of airplane in the First World War—were held together by struts and wires. A pilot would get into trouble if the turbulence-induced transition from high to low drag occurred during lift-off or landing. Therefore, struts with different cross section were investigated in the wind tunnel. The onset of turbulence in the boundary layer became thus a subject of experimental war research (Munk 1917).

---

[42]Prandtl (1914, p. 183). Translation ME.
[43]Prandtl (1914, p. 189). Translation ME.

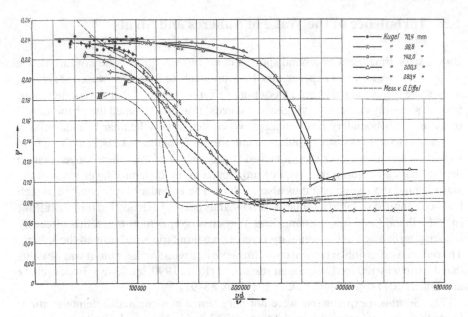

Fig. 1.3 Transition to turbulence in the boundary layer of spheres results in a drop of the coeffient of friction ψ (displayed as a function of the Reynolds number). The measurements from Eiffel's wind tunnel (dashed curves) indicate that the transition to turbulence occurred at a lower Reynolds number than in Prandtl's wind tunnel (solid curves). Prandtl explained this by a higher degree of turbulence in Eiffel's wind tunnel (Prandtl 1914, Fig. 1) (Courtesy University of Göttingen)

With a host of military requests for wind tunnel investigations related to all kinds of aerodynamic measurements (mainly concerning the lift and drag of airfoils) the theoretical analysis of this turbulence effect had to wait. In 1916 Prandtl penned for his future research a "Working program for a theory of turbulence" (Eckert 2006, Sect. 5.2). Like Blasius in Klein's seminar, Prandtl distinguished between the onset of turbulence and fully developed turbulence. We may discern here the root for Prandtl's turbulence research in the interwar period (Bodenschatz and Eckert 2011, pp. 47–52).

Chapter 2
The Turbulence Problem in the 1920s

Abstract In the early 1920s the turbulence problem was perceived as the quest for a theory concerning the onset of turbulence. Its solution was expected along the Orr-Sommerfeld approach. It reached centre stage as a research subject for applied mechanics and mathematics. By the mid 1920s, however, Prandtl regarded fully developed turbulence as the greater turbulence problem for which he suggested the "mixing length" concept. A test case was the turbulent friction along a wall. Empirical data suggested at first that the mean velocity profile with growing distance from the wall obeys a power law; by 1929 it became clear that for high Reynolds numbers it was rather a logarithmic law. The derivation of the "wall law" became subject of a fierce rivalry between Prandtl and Kármán.

In the early 20th century "the true content of the problem of turbulence," to resume Sommerfeld's view from his Rome paper in 1908, had been perceived as the quest to determine the limit of stability of laminar flow. Thus it was expected to bridge the gap between theoretical hydrodynamics and practical hydraulics. The advent of aeronautics shifted the focus on aerodynamics—a research field whose lack of theoretical underpinning became exposed by wind tunnel investigations shortly before and during WW I. After the war, fluid mechanics became the wider umbrella for investigations of flow problems disregarding their occurrence in water, air or other fluids. By the same token, turbulence became regarded as a universal phenomenon beyond the former association with hydrodynamics, hydraulics and aerodynamics.

2.1 The Turbulence Problem in *ZAMM*

The main difference between the turbulence problem as perceived before and after WW I was that it became declared as a challenge for a new scientific community. Before the war, the annual Naturforscher meetings provided the only opportunity for applied mathematicians, physicists, and engineers in German speaking countries to express common concerns. When they met again for the first time after the war in 1920 in Bad Nauheim, Prandtl and others who perceived their experience from

M. Eckert, *The Turbulence Problem*,
SpringerBriefs in History of Science and Technology,
https://doi.org/10.1007/978-3-030-31863-5_2

the war as a lesson for the future, expressed the need for a new umbrella for their research field. The result of their collaborative effort was the establishment of a new professional society, the Gesellschaft für angewandte Mathematik und Mechanik (GAMM), with Prandtl as president (Gericke 1972, pp. 5–10; Tobies 1982). Their mouthpiece was the *Zeitschrift für angewandte Mathematik und Mechanik (ZAMM)* founded shortly after the war and edited by Richard von Mises.

Already in the editorial to the first issue of *ZAMM* turbulence was declared as one of the themes within the scope of the journal, particularly problems about the circumstances responsible for the onset of turbulence (von Mises 1921, p. 12). For more details, Mises hinted at Fritz Noether's forthcoming paper on "The turbulence problem" assigned to the rubric of comprehensive review reports. "By the turbulence problem, generally speaking, we understand the question why Poiseuille flow, although it is always possible, is realized only within a restricted area and makes room for hydraulic flow." Thus Noether defined the turbulence problem once more as the key to bridge the gap between hydrodynamics and hydraulics. He even confined "the use of the term 'turbulent' only for a certain area of transition between laminar and hydraulic flow."[1] Noether's review portrayed the efforts to solve the turbulence problem by the method of small oscillations (i.e. what later was called the Orr-Sommerfeld approach) as a sequence of failures. The "energy method" introduced by Reynolds and Lorentz also appeared little promising. By and large, theory was in a dead end because it failed to predict a limit beyond which laminar flow would become unstable (Eckert 2010).

The same message was echoed from a different perspective in another *ZAMM*-paper on "Experimental investigations on the turbulence problem." Its author was Ludwig Schiller, an experimental physicist from Leipzig University. "Up to now the numerous theoretical investigations about the 'turbulence problem' have not yet led to a satisfying result", Schiller introduced his paper.[2] He had just spent a longer research sojourn with Prandtl in Göttingen where he performed exhaustive experiments on pipe flow. The results became the subject of his habilitation thesis about "Investigations on laminar and turbulent flow in circular pipes" which promoted him in 1921 as Privatdozent for "physics and aeronautics" and in 1926 as extraordinary professor for "mechanics and thermodynamics" at the physics institute of Leipzig University.[3] The allocations of Schiller's specialty foreshadow the ambivalence of academic physics with regard to applied mechanics—an ambivalence that would also affect future research of physicists on turbulence. For the time being, Schiller's *ZAMM*-paper provided evidence that the turbulence problem was not merely an issue for theorists. Based on the results of his habilitation he concluded that the critical Reynolds number for the transition from laminar to turbulent pipe flow could be as low as 200 and as large as 25500. For "technically smooth" pipes Schiller discerned the Reynolds number 1160 as a kind of true critical Reynolds number. Below this limit any vortices present in the inflow would always disappear within a sufficient

---

[1] Noether (1921, p. 126). Translation ME.
[2] Schiller (1921, p. 436). Translation ME.
[3] Personal File L. Schiller, PA 0254, University Archive Leipzig.

calming zone. To each Reynolds number above 1160 "a certain disturbance is necessary in order to elicit turbulence. The larger the Reynolds number the lower the disturbance that is required to this."[4]

Schiller presented the results of his habilitation thesis also in September 1921 in Jena at a common conference of the Deutsche Mathematiker-Vereinigung, the Deutsche Physikalische Gesellschaft und the Deutsche Gesellschaft für Technische Physik where Prandtl and the likes of him invited "scientific engineers" to showcase their research on applied subjects in order to foster the foundation of what became the GAMM (Eckert 2017, Sect. 5.5). The turbulence problem thus became instrumental for the formation of a new scientific community. Prandtl added his own "Remarks on the genesis of turbulence" to the debate. They were published both in the new *ZAMM* (Prandtl 1921) and in the *Physikalische Zeitschrift* (Prandtl 1922). Prandtl's remarks added to the turbulence problem a surprising aspect because "we found an instability of small oscillations contrary to the dogma." What Prandtl called "the dogma" was the conclusion of almost all previous studies that the Orr-Sommerfeld approach failed to yield a critical Reynolds number, i.e. "the dogma" predicted stability for all Reynolds numbers (Eckert 2010). Prandtl and his doctoral student, Oskar Tietjens, arrived at the opposite result. "We could not believe this result at first and repeated the computation independently from another in three different ways. But there was always the same sign which meant instability."[5]

Both "the dogma" and Prandtl's opposite conclusion, of course, were not compatible with the empirical evidence observed in real flows. Prandtl's presentation was followed by a vivid debate with Sommerfeld, Kármán and others (Prandtl 1922, pp. 5–6). Previously "the turbulence problem" was that all theories failed to yield a transition to turbulent flow up to the highest Reynolds numbers; now it was the opposite, that there should not even exist a laminar flow because the slightest disturbance would make it unstable even at the lowest Reynolds numbers. The presentations on "the turbulence problem" arose "great interest," a reviewer highlighted the debate in *ZAMM*, and the Jena meeting as a whole "completely fulfilled the expectation that here, for the first time, within the annual meetings of mathematicians and physicists applied mathematics and mechanics showed to advantage to a larger extent and with one accord."[6]

2.2 A New International Forum for Applied Mechanics

In August 1922, *ZAMM* announced for the following month an international "hydro-aerodynamic conference" in Innsbruck (Austria).[7] It was largely the result of a private initiative, conceived and carried through by Theodore von Kármán. After his return

---

[4]Schiller (1921, p. 443). Translation ME.
[5]Prandtl (1921, p. 434). Translation ME.
[6]*ZAMM* 1, 1921, pp. 419–420. Translation ME.
[7]*ZAMM* 2, 1922, p. 322.

from the war, Kármán was eager to establish at the Technische Hochschule Aachen a budding center for modern fluid mechanics along the role model of Prandtl's Göttingen institute. He shared with Prandtl also the ambition to foster applied mechanics as a rewarding challenge for scientific engineers. As he wrote in April 1922 to the mathematician Tullio Levi-Civita in Italy, "unfortunately the personal intercourse among those who work in this area is rather sparse." Thus he attempted to extend the initiative that had just resulted in the formation of *ZAMM* and GAMM beyond the national borders of Germany. He suggested to convene "a very unofficial meeting" at Innsbruck as a location which could be regarded so shortly after the war as "neutral soil". If "such a casual meeting" would be successful, more official international conferences could follow. He asked Levi-Civita to invite the colleagues from the Romance-language- and English-speaking countries, while he would care for those in Germany, the Netherlands, Austria, Switzerland, Russia, Czechoslovakia, and Scandinavia. He emphasized that he would extend the invitation to England and France, "but I cannot judge whether in these countries the friendly attitude is far enough advanced so that an invitation would not be rejected."[8]

Overcoming national resentments few years after the war proved difficult, but the success of the Innsbruck meeting added to the momentum to form a truly international community. Kármán and Johannes Martinus Burgers, the director of the Laboratory for Aero- and Hydrodynamics at the technical university in Delft, felt encouraged to plan a sequel mechanics congress with a more official character.[9] This congress was held in September 1924 in Delft. Altogether there were 214 participants from 21 countries—a remarkable demonstration of international scientific cooperation at a time when science was still devided into hostile camps between the former Central Powers and the Entente (Kevles 1971). *ZAMM* reported about this "first international congress for applied mechanics" enthusiastically and announced the resolution of the congress committee to organize the next such event in 1926 in Zurich and then every four years in another country.[10]

The international mechanics conferences also reflect a gradual change in the perception of the turbulence problem. At the Delft congress Kármán reported "On the stability of laminar flow and the theory of turbulence," the Norwegian meteorologist Halvor Solberg "On the turbulence problem" and the Russian mathematicians L. V. Keller and Alexander A. Friedmann on "Differential equations for the turbulent motion of a compressible fluid" Biezeno and Burgers (1924). As these titles suggest, the turbulence problem was by that time already perceived in a broader sense. At the Innsbruck meeting, for example, Werner Heisenberg, then a young student of Sommerfeld, had approached turbulence from a new vantage point by looking directly for "nonlaminar solutions" of the Navier-Stokes equations. "We ask therefore for turbulent motion itself, for its appearance and for the range of Reynolds numbers

---

[8] Kármán to Levi-Civita, 12 April 1922. TKC 18.8. Translation ME.

[9] Burgers to Kármán, 15 May 1923. TKC 4.21.

[10] *ZAMM* 3, 1924, pp. 272–276. For more details about the birth of the international mechanics congresses see Battimelli (1988) and Eckert (2006, Sect. 4.3).

R for which it is possible."[11] Two years later the turbulence problem became the subject of his doctoral dissertation[12]:

> For our purpose it is sufficient to provide a very rough sketch about the present state of the turbulence problem. The investigations so far may be categorized in two groups: the investigations of the first group deal with the stability analysis of some laminar motion, the other with turbulent motion itself.

Thus the turbulence problem was no longer confined to the onset of turbulence only. Schiller shared this broader view when he argued that "almost all fluid motions that appear in practice or motions of solid bodies in air or in fluids belong to the 'turbulence problem'. This wider definition contrasts with the relatively narrow meaning of the 'turbulence problem' in the mathematical-physical literature", which should rather be called "the stability problem of hydrodynamics". But he observed that among theorists also a wider notion was adopted. "Nowadays one has extended in the mathematically oriented literature the notion of the 'turbulence problem' insofar as one has added to the former problem that of fully developed turbulent motion."[13]

2.3 The "Great Problem of Developed Turbulence"

The tendency to extend the notion of the "turbulence problem" became most apparent when Prandtl reported by the same time in *ZAMM* about a new approach with which he attempted "to compute the distribution of the main flow in turbulent motion under various conditions hydrodynamically."[14] This approach became the subject of his presentation at the forthcoming international mechanics congress held in 1926 in Zurich. There Prandtl focused on "what I should like to call the 'great problem of developed turbulence', a deeper understanding and a quantitative computation of processes by which existing eddies despite their damping by friction create over and over new ones". But he expected that this problem "will perhaps not so soon be solved."[15]

Prandtl's approach became known as the mixing length concept. It was based on the introduction of a characterisic length which played a similar role for turbulence as the free path length in the kinetic theory of gases. Prandtl interpreted this length as the path that an eddy moves through the turbulent flow until it looses its identity by mixing with neighbouring parts of the fluid. "According to this meaning we call it mixing length and designate it by l." He assumed that l is a mean length of motion in direction y perpendicular to the main flow in x direction with velocity $\bar{u}$, and that in first approximation the velocity of an eddy in this transversal motion has a

---

[11] Heisenberg (1922, p. 139). Translation ME.
[12] Heisenberg (1924, p. 577). Translation ME.
[13] Schiller (1925, pp. 566–567). Translation ME.
[14] Prandtl (1925, p. 137). Translation ME.
[15] Prandtl (1927, p. 62). Translation ME.

velocity that equals $l \frac{\partial \bar{u}}{\partial y}$. Using this expression in Boussinesq's eddy viscosity— by dimension the product of a velocity with a length—yielded the turbulent shear stress in terms of l (Prandtl 1927, pp. 62–64).

In contrast to earlier attempts to account for turbulence in terms of Boussinesq's eddy viscosity (see, e.g., Hahn et al. 1904, pp. 62–64), l could be adjusted to the boundary conditions of a specific turbulent flow and in general be assumed to be a function of spatial variables such as the distance from the wall in pipe or channel flow. In these cases, however, the approach met with problems (see Sect. 2.5). Therefore, Prandtl used at first an example of "free turbulence" without walls, such as the broadening of a turbulent jet ejected in an ambient fluid at rest. In this case he assumed that the mixing length increases proportional to the width of the jet in each cross-section behind the nozzle. He assigned the computation to his student Walter Tollmien who arrived at results in excellent agreement with experimental measurements. Thus the mixing length offered at least in one case a viable route to solve the turbulence problem "hydrodynamically", although this did not mean that the theory was based on first principles only. The crucial assumption about the mixing length could only be made plausible retrospectively by comparison with experiments.[16]

Prandtl's mixing length concept was the result of that part of his "Working program for a theory of turbulence" which was dedicated to fully developed turbulence (see Sect. 1.5). He had penned it ten years ago in the midst of WW I, but put on the agenda of his institute only "about five years ago", as he mentioned in the introduction to his Zurich lecture. Furthermore, despite the goal of "a theory of turbulence," the program involved experiments. At Zurich he presented photographs and even a film that showed the turbulent flow in a channel. His expectation that the problem "will perhaps not so soon be solved" was not the least due to these experiments: "our photographic and cinematographic records only show how hopelessly complex these motions are even in the case of smaller Reynolds numbers."[17]

2.4   Tollmien's Solution of the "Stability Problem"

While Prandtl was aiming with the mixing length concept at the "great problem of developed turbulence," he did not loose sight of the other part of the turbulence problem concerned with the onset of turbulence, which Schiller had named more appropriately the "stability problem". Prandtl's presentation at the Jena meeting in 1921 and the detailed elaboration by his doctoral student Oskar Tietjens in *ZAMM* (Tietjens 1925) did not seem to offer a loophole through which the deadlock of the Orr-Sommerfeld approach could be overcome. But Prandtl must have regarded

---

[16]The theory of jet broadening was published in *ZAMM* (Tollmien 1926). For more details on the genesis of the mixing length approach see Eckert (2006, Sect. 5.3) and Bodenschatz and Eckert (2011, pp. 54–56).

[17]Prandtl (1927, p. 62). Translation ME.

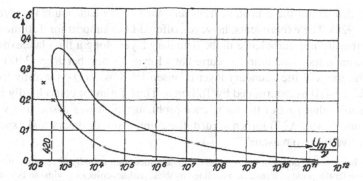

Fig. 2.1 Tollmien's indifference diagram displays the demarcation between stable and unstable states of flow (Tollmien 1929, Abb. 4) (Courtesy University of Göttingen)

Tollmien's successful application of the mixing length concept to turbulent jet broadening as a recommendation to assign Tollmien now the stability problem as subject of his doctoral work.

Tollmien titled his dissertation "On the origin of turbulence" (Tollmien 1929). Instead of the piecewise linear velocity profile which had doomed Tietjens's analysis to failure, Tollmien chose for the main flow a profile without kinks, close to that which Prandtl and Blasius had predicted for the laminar boundary layer along a flat plate. After tedious mathematical manipulations he arrived at the long-sought distribution of stable and unstable states of flow. "We note at first that an extraordinarily narrow range of oscillations becomes dangerous for the laminar flow," he remarked about the shape of the indifference curve which marked the limits of stability. "In the same way as there is a lower limit of 420 for the Reynolds number there is an upper limit for the disturbance parameter beyond which there is no instability."[18] The "disturbance parameter" was proportional to the wave number of the disturbance (Fig. 2.1). In other words, the instability which preceded the onset of turbulence depended both on Reynolds number and disturbance in a way that made it difficult to discern a clear-cut origin of turbulence.

From the perspective of experiments the laminar boundary layer along a flat plate was still virgin ground. Prior to the mid-1920s there were no comparable data about the transition to turbulence in a boundary layer like those of Saph and Schoder or Blasius in pipe flow. With the rise of aeronautical laboratories, however, measurements in the air steam of wind tunnels added to those of water flow in pipes and channels in hydraulic laboratories. At the international mechanics congress in 1924 in Delft, Burgers presented precision measurements about the velocity profile along a flat plate, the subject of his doctoral student's B. G. van der Hegge Zijnen experiments with the novel hot-wire technique in a wind tunnel (Burgers 1925). Four years later, M. Hansen published similar results from Pitot-tube measurements in a wind tunnel in Kármán's Aerodynamic Laboratory at the Technische Hochschule Aachen,

---

[18]Tollmien (1929, p. 42). Translation ME.

including data about the transition from laminar to turbulent boundary layer flow (Hansen 1928). The experiments, however, offered little support for Tollmien's theory. The transition to turbulence in the boundary layer along a flat plate exposed to the airstream in the wind tunnel occurred at a Reynolds number $R_\delta \sim 3100$, where δ is the thickness of the boundary layer (Hansen 1928, p. 193)—far away from the lower limit of 420 as determined by Tollmien. Thus Tollmien could hardly pretend to have solved that part of the turbulence problem which was associated with the onset of turbulence. As Tollmien argued, the comparison of theory and experiment was "abortive for two reasons"[19]:

> Firstly, because one does not know about the disturbances that actually occur; then it is not clear how far the point of transition as defined by these authors coincides with our beginning of instability of the laminar flow. Burgers, and similarly Hansen, define the point of transition in this way: In the front part of the plate there is a laminar flow which develops according to Prandtl-Blasius; beyond the point of transition there is a pronounced turbulent part whose laws have been elucidated by Prandtl and von Kármán (1/7 power law etc).

With the latter remark Tollmien alluded to an ongoing rivalry between Prandtl and Kármán about the theory of the turbulent boundary layer which became known as the quest for a "universal law of turbulence".

2.5   The Quest for a Universal Law of Turbulence

The rivalry had begun in the early 1920s. "Dear Master, colleague, and former boss," Kármán had introduced in 1921 a five-page letter to Prandtl in which he sketched "a kind of 'turbulent boundary layer theory'."[20] He derived the velocity profile in turbulent wall-bounded flows, $U \sim y^{1/7}$, where U is the mean velocity parallel to a flat plate and y the distance perpendicular to the surface of the plate. From the published version of the theory (von Kármán 1921) it becomes clear why he designated his considerations only as "a kind of 'turbulent boundary layer theory'." The 1/7th power law was not derived from first principles but based on the empirical law of turbulent pipe flow established by Blasius in 1913, whereupon the loss of pressure per unit length varies like $U^{7/4}$ (see Sect. 1.4). The derivation started from Blasius empirical formula and employed dimensional analysis in order to extrapolate from pipe flow to the velocity profile on a flat plate.

Prandtl had arrived at the same result earlier without publishing it. He had mentioned it in informal talks, as Kármán recalled in his letter. Prandtl's derivation of the 1/7th law appeared in print only five years later in the doctoral dissertation of Johann Nikuradse to whom Prandtl had assigned experimental investigations of turbulent flow in a water channel. Nikuradse's measurements agreed with the 1/7th law (Nikuradse 1926), but this corroboration contributed little to the theory of the turbulent boundary layer; it merely confirmed Blasius's empirical formula upon which

---

[19]Tollmien (1929, p. 43). Translation ME.

[20]Kármán to Prandtl, 12 February 1921. GOAR 3684. Translation ME.

the 1/7th law was based. "You ask for the theoretical derivation of Blasius' law for pipe friction," Prandtl once responded to the question of a colleague. "The one who will find it will thereby become a famous man!"[21]

When Prandtl developed by the same time his mixing length approach he hoped that it would serve as a new starting point for the theory of the turbulent boundary. However, he failed to derive the 1/7th law from one or another plausible assumption for the mixing length. He ruled out a linear relation between mixing length and distance from the wall, like in the case of the broadening of a turbulent jet (see Sect. 2.3), because this would have resulted in a logarithmic law with a singularity at the wall. Prandtl therefore dismissed the most plausible linear relation for the mixing length. In a lecture delivered in October 1929 in Tokyo during a trip around the world he argued: "The approach l proportional y does not lead to the desired result because it would yield $\bar{u}$ prop. log y."[22] By the same time, however, Nikuradse concluded from new measurements on pipe flow at very high Reynolds numbers that a logarithmic formula yielded indeed a better agreement with the data than a power law (Nikuradse 1930).

In the meantime, Kármán responded to the quest for a theory of wall turbulence from a new angle. In 1928, his doctoral student Walter Fritsch published the results of experimental investigations about turbulent channel flow with different wall surfaces; they showed that the velocity profiles could be superimposed upon one another if the shear stress at the wall was the same, regardless of the roughness of the wall surface (Fritsch 1928). From this observation, Kármán concluded that at some distance from a wall the velocity fluctuations are similar anywhere and anytime in fully developed turbulent flow. Therefore, he regarded the mixing length as a characteristic scale of the fluctuating velocities. "The only important constant thereby is the proportionality factor in the vicinity of the wall," he announced his theory in a letter to Burgers.[23] Instead of the 1/7th law, Kármán's similarity approach yielded a logarithmic wall law (von Karman 1930, 1931). The singularity did not bother Kármán because his approach was not valid immediately at the wall. Nevertheless he regarded his theory as universal because the only constant that entered his approach was independent of the experimental conditions. It was soon named after him the "Kármán constant".

Prandtl swiftly confirmed Kármán's results. He published his analysis in 1932 as a corollary to the presentation of new data in the series of his laboratory communications, *Ergebnisse der Aerodynamischen Versuchsanstalt zu Göttingen*. He duly acknowledged Kármán's earlier publications from the year 1930, but he claimed that he had arrived at the same results "at a time when Kármán's papers had not yet been known, so that once more, like ten years ago with the same problem, the thoughts in Aachen and Göttingen followed parallel paths".[24] In Kármán's recollection, however, Prandtl was "crestfallen" and "chagrined" that his former pupil had once more succeeded "in exercising his well-known talent for skimming the cream off the milk."

---

[21] Prandtl to Birnbaum, 7 June 1923. MPGA, Abt. III, Rep. 61, Nr. 137. Translation ME.
[22] Prandtl (1930, p. 9). Translation ME.
[23] Kármán to Burgers, 12 December 1929. TKC 4.22. Translation ME.
[24] Prandtl (1932, p. 21). Translation ME.

The quest for a "universal law of turbulence" had turned into a "first-class rivalry" (von Kármán 1967, pp. 135–138).

According to the dates of Kármán's and Prandtl's publications it is obvious that Kármán was the winner in this race. As we learn from their contemporary correspondence, however, the rivalry was not just about priority. The bone of contention became a report in the journal *Werft, Reederei, Hafen* about a conference at the Hamburgische Schiffbau-Versuchsanstalt where Prandtl's results were presented as the authoritative state of the art. In a normal priority dispute it would have been sufficient for Kármán to point to his publications from the year 1930 in order to correct this erroneous view. But "who reads G. N. (Göttinger Nachrichten) and Stockholm Kongress," Kármán complained in a letter to Prandtl; he insisted that Prandtl acknowledged his contribution in the *Ergebnisse der Aerodynamischen Versuchsanstalt zu Göttingen* as the "standard for practitioners."[25] Kármán wished that his achievement was regarded as a milestone to practical engineering rather than pure science! Both Prandtl and Kármán regarded the "practitioners" as their true peers, i.e. the engineers of hydraulic and aeronautical laboratories. In other words: They perceived their approaches towards the turbulence problem, such as the derivation of the "universal law of turbulence" for wall-bounded flow, first and foremost as contributions to engineering rather than as solutions of the last riddle of classical mechanics.

---

[25] Kármán to Prandtl, 26 September 1932. AMPG, Abt. III, Rep. 61, Nr. 793. Translation ME. For more detail on this rivalry see Eckert (2017, Sect. 6.8).

Chapter 3
The Rise of Statistical Theories of Turbulence

Abstract In the mid-1930s the turbulence problem became the subject of a statistical theory developed by Geoffrey Ingram Taylor. It was to a large extent motivated by the concern about wind tunnel turbulence. From an experimental perspective the breakthrough came with the hot-wire technique that enabled precise measurements of grid-turbulence in wind tunnels. Theodore von Kármán and Leslie Howarth extended Taylor's theory. In 1938, statistical theories of turbulence took centre stage at a symposium on turbulence as part of the Fifth International Congress for Applied Mechanics. Apart from the mainstream statistical theories, Johannes Martinus Burgers developed simplified mathematical models in an attempt to capture the principal features of turbulence—with an alternative view on statistical theories.

Theodore von Kármán's recent "examination of the turbulence problem" appeared to Fritz Noether as "appropriate to bring this still unsolved riddle closer to its elucidation." Noether referred to the similarity hypothesis in Kármán's theory of fully turbulent channel flow (von Kármán 1930) and expected that along this way "statistics can be applied in a precise manner to the turbulence problem."[1] This was not the first time to call for a statistical theory as a solution of the problem of fully developed turbulence. Kármán had argued already in 1921 that fully turbulent flow might be controlled by some unknown, statistical law, and one or another such expectation was raised by Richard von Mises, Prandtl and Burgers throughout the 1920s (Battimelli 1984).

Some statistical arguments had always been involved in the theory of fully developed turbulence, such as in the concept of Boussinesq's eddy viscosity or Reynolds's derivation of averaged Navier-Stokes equations. When Prandtl had penned in 1916 his working program for a theory of turbulence he noted under the headline "Developed turbulence" that one should assume for wall-bounded flow a "statistical equilibrium of a mass of vortices in the vicinity of a wall" (Eckert 2006, Sect. 5.2). Prandtl's mixing length was supposed to be a kind of mean free path length for turbulent eddies. Yet the omnipresence of statistical concepts had little impact on the perception of the

---

[1] Noether (1931). Translation ME.

turbulence problem. This changed in the mid-1930s when Geoffrey Ingram Taylor published a series of papers that contain the seeds for all subsequent work on the statistical theory of turbulence (Sreenivasan 2011, Sect. 4.5). The statistical notions developed by Taylor and his successors were considered essential whenever the turbulence problem was at stake.

3.1 Atmospheric Turbulence

Prior to his theory from the mid 1930s Taylor had attempted to develop a statistical approach from observations of atmospheric turbulence. Meteorological investigations during WW I had suggested "that turbulent motion is capable of diffusing heat and other diffusible properties through the interior of a fluid in much the same way that molecular agitation gives rise to molecular diffusion" (Taylor 1921, p. 196). Unlike molecular diffusion, the subject of the kinetic theory of gases, diffusion by turbulence had not yet become subject of theoretical investigations. The problem was suited for the so-called Lagrangian description where the path of individual particles is followed instead of the Eulerian description where the particles are registered by their position at fixed points in space. Taylor attempted to determine the (Lagrangian) correlation function of a fluid element as a function of time. He arrived at a formula for the mean-square displacement in one direction (similar to the description of a particle in Brownian motion) which he used to predict the diffusion of smoke emitted from a fixed point in a wind—as described in experiments by the meteorologist Lewis Fry Richardson. The theory also applied to the conduction of heat in a turbulent flow. The results showed "why the 'diffusing power' of any type of turbulence appears to depend so little on the molecular conductivity and viscosity of the fluid" (Taylor 1921, p. 196).

Unlike Noether and others who associated the turbulence problem with the onset of turbulence, Taylor did not advertise his statistical theory on turbulent diffusion as an effort to solve the turbulence problem. His focus was more on practical applications, particularly "The eddy motion of the atmosphere," as he had titled an earlier paper (Taylor 1915). He shared this orientation with Richardson who derived from observations of atmospheric turbulence the idea of what became famous as "Richardson cascade":

big whirls have little whirls
that feed on their velocity,
and little whirls have lesser whirls
and so on to viscosity

The rhyme is often presented like a prophesy of Kolmogorov's theory from the year 1941 (see Sect. 4.1). However, the original quote sounds more prosaic. It is completed with "in the molecular sense" (which distorts the rhyme) and preceded by observations of pilots from contemporary airplanes: "The upward currents of large cumuli give rise to much turbulence within, below, and around the clouds, and the structure of the clouds is often very complex." Similarly, Richardson added his frustration "when making a drawing of a rising cumulus from a fixed point; the details change before the sketch can be completed. We realize thus that: big whirls have little whirls that feed on their velocity, and little whirls have lesser whirls and so on to viscosity—in the molecular sense" (Richardson 1922, p. 66).

A few years later, Richardson made another attempt to cope with the peculiar features of atmospheric turbulence. His starting point was "Fick's equation" of diffusion with the "so-called constant K" of diffusivity—which varied "in a ratio of 2 to a billion" when applied to meteorological measurements. The goal of Richardson's study was "to comprehend all this range of diffusivity in one coherent scheme" (Richardson 1926, p. 709). Richardson suggested a "non-Fickian" diffusion equation which involved a scale-dependent diffusivity "$F(l) = \varepsilon l^{4/3}$", where the scale l is the root mean square displacement of a pair of particles and ε a constant which had to be determined from observational data.

With hindsight, Richardson's 4/3-law and his cascade of whirls antedated Kolmogorov's theory (see Benzi (2011) for a modern perspective of Richardson's achievements). Nevertheless Richardson did not envisage an all-embracing statistical theory of turbulence. Even the word turbulence appears only twice in his 1926 paper, and both times associated with atmospheric observations. By and large, the development of the statistical theory of turbulence had to await other impulses that enabled a more reliable interplay between theory and experiment than the weather.

3.2 Wind Tunnel Turbulence

Such impulses came from wind tunnel measurements. The rational argument of all model testing in wind tunnels—that the coefficients of lift and drag measured in a wind tunnel and those in free flight are the same if the model is geometrically similar and the Reynolds number is the same—ignored turbulence. Since Eiffel's and Prandtl's experiments with spheres in the wind tunnel it was clear that turbulence affected model testing (see Sect. 1.5). Hugh Dryden, the director of the Aerodynamical Physics Section at the National Bureau of Standards in Washington, asked Prandtl in 1921 about "your idea as to the physical conception of the turbulence" because "the most important wind tunnel problem is a study of turbulence and its effects."[2]

Since the 1920s aerodynamic laboratories dedicated a considerable research effort to wind tunnel turbulence. The degree of turbulence was modified by screens of different mesh size in a wind tunnel. In a study at the MIT it was found "that the turbulence tends to die out more rapidly downstream as the screen becomes finer" (LePage and Nichols 1924, p. 5). But it was difficult to measure the degree of turbulence. In an effort to standardize wind tunnel tests engineers from the NACA compared the drag measurements on spheres obtained in various wind tunnels (Bacon and Reid 1924).

---

[2]Dryden to Prandtl, 5 March 1921. MPGA, Abt. III, Rep. 61, Nr. 361. For more detail see Eckert (2006, Sect. 5.5).

"The data collected here must be considered, primarily, as data concerning the tunnel, and not the models tested there," another study concluded in 1925. In other words, wind tunnel turbulence cast doubt on the reliability of model testing. "The actual process of standardization still lies in the future" (Reid 1925, p. 219).

In order to study the turbulence in wind tunnels more closely, Dryden and his collaborators at the National Bureau of Standards performed a serious of measurements about the resistance of cylinders at different degrees of turbulence (caused by different wire meshes in the 54-inch tunnel of their laboratory). "The investigation was planned along two lines, the problem of turbulence being attacked in two directions," they argued about their procedure. The first aimed at the drag of the cylinders, the second at pressure variations measured by Pitot tubes. They reported some success about recording fluctuations, but their ambitions reached beyond that. "From a more fundamental point of view we can never be satisfied until we know more definitely the nature of turbulence; in other words, its cause and the exact manner in which it manifests itself as changes in the air speed, direction, and pressure" (Dryden and Heald 1926, p. 465).

The next step towards this goal was to employ hot-wire anemometry for measuring the fluctuating velocity of the air stream in the wind tunnel. The technique was known for several years and occasionally applied in the wind tunnel, but the sophisticated electronics required for translating temperature changes (due to the cooling of the hot wire by the air stream) into electrical signals made it cumbersome to handle. The apparatus in Dryden's laboratory was "very bulky, far from portable, and in many respects inconvenient to use" (Dryden and Kuethe 1929b, p. 25). As soon as the hot-wire anemometer was put into service Dryden and his collaborator suggested a definition for the degree of turbulence in a wind tunnel: "The turbulence at a given point is taken to be the ratio of the square root of the mean square of the deviations of the speed from its mean value to the mean value. The turbulence is a mean fluctuation taken in a definite manner and expressed as a percentage of the mean speed" (Dryden and Kuethe 1929a, p. 8). They illustrated their definition with a diagram that showed how the critical Reynolds number of spheres (at which the drag coefficient drops to the lower value) depends on the so-defined percentage of wind tunnel turbulence. They suggested to compare different wind tunnels by assigning a characteristic number to each tunnel so that wind tunnels with the same number yield comparable results of model tests. "This characteristic number will be the measured turbulence or, more conveniently, the Reynolds Number for which the sphere drag is 0.3" (Dryden and Kuethe 1929a, p. 22).

The hot-wire apparatus at the National Bureau of Standards opened a new era of turbulence measurements. Yet its cumbersome use did not immediately prompt new attacks of the turbulence problem. "The great need at the present time in the further development of the theory of turbulence is more experiments on the actual fluctuation to supplement the data already available on the distribution of mean velocity." Thus Dryden's group assigned in 1929 priority to experiments (Dryden and Kuethe 1929b, p. 5). Within few years they further improved the hot-wire technique (Mock and Dryden 1932). By 1931 they were able to measure "any desired characteristic of the

turbulent flow," Dryden informed Prandtl about their progress, "perhaps only with
great difficulty and complicated apparatus, but surely and with reasonable accuracy."[3]

3.3 Taylor's and Kármán's Statistical Theories

By the early 1930s measurements of wind tunnel turbulence with hot-wire anemometry were also performed at the National Physical Laboratory (NPL) in England
and in Prandtl's institute in Göttingen. "We ourselves have conducted an experiment
in which two hot-wires are placed at larger or smaller distances from each other,"
Prandtl informed Taylor about an experiment in a wind tunnel with a rectangular
cross section (100 cm x 25 cm) which aimed at the measurement of correlations of
turbulent velocity fluctuations at different positions (Prandtl and Reichardt 1934). "In
any case, I am as convinced as you that from the study of those correlations as well
as between the direction and magnitude fluctuations, for which we have prepared a
hot-wire setup, very important insights into turbulent flows can be gained."[4] At the
request of and in collaboration with Taylor, the diffusion of turbulence downstream
from an eddy generating grid was measured with hot-wire anemometry also in a
one foot wind tunnel at the NPL (Simmons and Salter 1934). In 1935 Taylor started
to publish his "Statistical theory of turbulence" in five parts in the *Proceedings of the
Royal Society London* (Taylor 1935a, b, c, d, 1936). It was based on the assumption
that the average values of the turbulent velocity fluctuations were isotropic, i.e. the
same in all directions— which appeared legitimate for grid-generated turbulence in
wind tunnels, where at some distance from the grid an observer moving downstream
with the mean velocity of the airstream would see in all directions the same average
fluctuations. At the core of Taylor's theory was the concept of a characteristic length
scale λ defined in terms of the correlation between neighbouring velocity fluctuations. It could be perceived as "the average size of the smallest eddies" (Taylor 1935a,
p. 421), or, in the eyes of the practitioner concerned with wind tunnel turbulence, the
"scale of turbulence" (Dryden et al. 1936, pp. 109).[5]

With regard to the turbulence problem it is interesting to see how Taylor and
those with whom he exchanged opinions about his work rated this theory at the time.
Taylor avoided the rhetoric of "the turbulence problem" but rather emphasized the
practical uses of his theory for wind tunnel turbulence. A few months before he began
to publish the first part of his theory he informed Prandtl that he had succeeded in
deriving "two formulae which seem to have practical interest."[6] One concerned the
decay of wind tunnel turbulence behind a grid (Taylor 1935a):

---

[3]Dryden to Prandtl, 8 May 1931. MPGA, Abt. III, Rep. 61, No. 361.

[4]Prandtl to Taylor, 23 December 1932. MPGA, Abt. III, Rep. 61, Nr. 1653. Translation ME.

[5]An appraisal of Taylor's theory is beyond the scope of this brief history. For a short review and
hindsight evaluation see Sreenivasan (2011, Sect. 4.5).

[6]Taylor to Prandtl, 2 March 1935. MPGA, Abt. III, Rep. 61, Nr. 1654.

$$\frac{U}{u'} = \frac{5x}{A^2 M} + const.$$

where U is the uniform velocity of the airstream in a wind tunnel, u' is the mean velocity fluctuation at a distance x behind a grid with mesh-size M, and A is a constant which is the same for grids of the same type.

The other practical result accounted for the critical Reynolds number of a sphere at which the drag coefficient suddenly drops from higher to lower values. When Dryden and his collaborators reviewed in 1936 their research on wind tunnel turbulence they interpreted their measurements in terms of Taylor's theory and explicitly acknowledged Taylor's involvement (Dryden et al. 1936, p. 110). Taylor's theory predicted that the critical Reynolds number depends on

$$\frac{u'}{U}(\frac{D}{M})^{1/5}$$

where D is the diameter of the sphere (Taylor 1936). "Except for the measurements made at a distance of 1 foot, the observations for both spheres and all screens lie remarkably well on a single curve, certainly within the observational errors," Dryden corroborated this result (Dryden et al. 1936, p. 124).

Taylor's theory also became the subject of discussions at the Fifth Annual Meeting of the American Institute of the Aeronautical Sciences (IAS) on January 27, 1937. Kármán presented at this occasion a theory which he was elaborating together with Leslie Howarth, a research fellow from the University of Cambridge who had come to Caltech in the summer of 1936 for a one-year sojourn. Taylor's statistical theory "raised considerable interest," Kármán introduced his presentation, "because it is concerned with the important problem of wind-tunnel turbulence and its results could be compared directly with experimental work done by Dryden in this country and by Fage, Townend and Simmons in England." Statistical concepts such as correlations had been introduced earlier, but Taylor's predecessors like Alexander Friedmann "could not carry his investigations to practical results" (von Kármán 1937a, p. 131). Dryden used this opportunity to suggest another theory focussing on the intensity and scale of wind tunnel turbulence as the crucial experimentally observed quantities for a statistical theory (Dryden 1937).

A few months later Kármán was honoured by the Royal Aeronautical Society in Great Britain to present the 25th Wilbur Wright Memorial Lecture. Kármán titled his talk simply "Turbulence". Once more, he exposed "the turbulence problem" as a highly competitive challenge. "Practically every scientist who has become interested in the turbulence problem has formed his own theory," Kármán portrayed the quest as an ongoing rivalry. "Because I, myself, have a theory in the making, I shall refrain from a criticism of the existing theories." But he was eager to demonstrate that this research was pertinent for aeronautical engineering (von Kármán 1937b, p. 1141):

> Many engineers to-day may consider the problem of turbulence merely as an interesting chapter of mathematical physics. They may be right. However, they should remember that if we meet a practical question in aerodynamic design which we are unable to answer, the reason that we are unable to give a definite answer is almost certainly that it involves

turbulence. Hence, I believe that in spite of the complex mathematical and physical aspect of the problem of turbulence, the scientist is justified in saying to the practical engineer: Tua res agitur (your case is on trial). It was the principal aim of my lecture to show this.

The Kármán-Howarth-theory focused on the correlations between turbulent velocity fluctuations at different locations and directions. It predicted, for example, a relation of the correlation coefficients $R_1(r)$ and $R_2(r)$ between fluctuating velocities at points separated by r and oriented in longitudinal and perpendicular direction of $\mathbf{r}$, respectively (von Kármán 1937a, p. 132):

$$r\frac{dR_1}{dr} + 2(R_1 - R_2) = 0.$$

"On reading this result the author realized that he had in his possession the material for submitting this relationship to experimental test," Taylor reacted to the Kármán-Howarth-theory. He referred to Simmons's recent wind tunnel measurements at the National Physical Laboratory in which turbulence was generated with a quadratic grid of 3 inches mesh-size. The formula represented "a true relationship for the turbulence produced in a wind tunnel by a grid" (Taylor 1937, pp. 312–313).

Yet Kármán and Taylor did not agree about the underlying assumptions of their theories. Kármán "has extended the author's statistical theory of isotropic turbulence in several directions," this was the way how Taylor referred to Kármán's theory (Taylor 1937, pp. 311). Kármán, however, regarded his theory as far more than a mere extension. It was more general and included "Taylor's consideration as a special case" (von Kármán 1937a, p. 131). As Kármán and Howarth noted in their comprehensive paper, some of their findings "are identical with Taylor's results"—but they attributed Taylor's achievement to his "remarkable vision for the relations between the quantities involved," whereas their results were derived from a coherent mathematical formalism. Nevertheless they were conscious about the lack of a physical underpinning. „It appears that the next step in the development of the theory should be to find the physical mechanism which is behind the mathematical relations" (von Kármán and Howarth 1938, p. 214).[7]

3.4   A Symposium on Turbulence

In September 1938 Taylor introduced a symposium on turbulence, organized as part of the Fifth International Congress for Applied Mechanics in Cambridge, Massachusetts, with a general lecture on "Some Recent Developments in the Study of Turbulence." He reminded the participants that turbulence had formed the subject of a general lecture at each one of the four preceding international mechanics congresses (Taylor 1939, p. 294). At the fourth congress in 1934 in Cambridge, England,

---

[7]The dispute is reflected in the correspondence between Kármán and Taylor by "sometimes testy exchanges", but otherwise did not affect their friendly relationship. See Leonard and Peters (2011, pp. 115–119) and Sreenivasan (2011, pp. 159–161).

it was planned to organize at the next Congress a special symposium on turbulence. As Jerome Hunsaker and Kármán emphasized in their capacity as members of the Congress Committee, the symposium on turbulence was supposed to be "not only the principal feature of this Congress, but perhaps the Congress activity that will materially affect the orientation of future research" (Hartog and Peters 1939, p. xvii). With Prandtl as chairman, Taylor as general lecturer and contributions by Kármán, Dryden and others, the symposium would also reflect what leading experts regarded as the turbulence problem.

In view of the preceding general lectures on turbulence, Taylor felt entitled to narrow the scope of his lecture in 1938 to the statistical theory of grid-generated wind tunnel turbulence. He also addressed the second part of the turbulence problem, the transition to turbulence, but he dismissed the instability theory as developed by Tollmien and Schlichting—not the least because his statistical theory suggested another cause for the onset of turbulence in the boundary layer: He argued "that turbulence acts as a boundary layer disturber through the medium of fluctuating local pressure gradients." As evidence he presented the formula which related the critical Reynolds number of a sphere in a wind tunnel to the degree of grid-generated turbulence from which the impact of the "disturber", i.e. the turbulent velocity fluctuations, on the transition to turbulence in the boundary layer of the sphere could be estimated.

Dryden presented in his contribution to the turbulence symposium a review of the experimental investigations of his group at the National Bureau of Standards. "Since spheres have been much used as indicators of turbulence, a comprehensive study was made of the effect of varying the intensity and scale of turbulence on the critical Reynolds Number of spheres," Dryden referred to these investigations on wind tunnel turbulence. "The experimental results were in good agreement with the relation suggested by G. I. Taylor" (Dryden 1939, pp. 362–363). Other parts of his review concerned the spectrum and decay of isotropic turbulence. As "second aspect of the general turbulence problem" he addressed studies on "approximately isotropic turbulence" and as third aspect the "non-isotropic turbulence in the approximately two-dimensional flow in the boundary layer of a plate." The latter would become a major research effort during World War II and result in the confirmation of the Tollmien-Schlichting theory (see Sect. 4.2). For the time being he could not yet report results because these investigations were either "in progress" or reserved for the "immediate future" (Dryden 1939, pp. 366–367).

By and large, the turbulence problem at this symposium was associated with fully developed turbulence. Kármán contributed "Some Remarks on the Statistical Theory of Turbulence," Norbert Wiener lectured on "The Use of Statistical Theory in the Study of Turbulence," Eric Reissner added a "Note on the Statistical Theory of Turbulence" – to name only some contributions explicitly devoted to statistical approaches (Hartog and Peters 1939). Only one participant, John L. Synge, an applied mathematician from the university of Toronto, dedicated his contribution to "The Stability of Plane Poiseuille Motion" (Synge 1939). Synge published in the same year an address on "Hydrodynamic Stability" for the Semicentennial of the American Mathematical Society, where he introduced his subject with the remark, that it "presents mathematical problems of no small difficulty: triumphs are few and disappointments many" (Synge 1938, p. 227). His address marks the departure of hydrodynamic stability theory as a research field in its own right – with ramifications to the onset of turbulence, but no longer as the key to the turbulence problem like twenty years ago (see Sect. 2.1).

If there was a common trend among the various contributions to the turbulence symposium, besides the emphasis on statistical approaches, it was an effort to establish a connex between theory and practice. Yet there was little agreement on the priorities of turbulence research. With hindsight one may observe seeds for future breakthroughs, like in Prandtl's contribution (Bodenschatz and Eckert 2011, p. 70), but from a contemporary perspective the riddles of turbulence appeared as inaccessible as ever. In Prandtl's view the symposium on turbulence was "not particularly successful." He acknowledged that "there have been gathered from all sides a whole lot of bricks and other raw materials, but it has not yet become apparent how the edifice should look like which can be built from it. Perhaps one has to prepare later a second symposium in which the construction plan can be clarified."[8]

Another critic was Johannes Martinus Burgers. Since the 1920s Burgers had attempted to tackle the turbulence problem by statistical methods without success (Alkemade 1995, Sect. E.4.2). As he recalled many years later, he had for some time "played with the hypothesis that a statistical theory of turbulence might be built upon the example of the statistical theory used in the kinetic theory of gases or in other conservative systems, provided the condition of constant energy content was replaced by a dissipation condition." This effort nourished his conviction that a theory of turbulence should first of all account for the basic difference between dissipative and conservative systems. "I thought that it would be necessary therefore to study the behaviour of dissipative systems, and that since the Navier-Stokes equations are so refractory, it might be helpful to replace them by a more elementary equation" (Burgers 1975, p. 9).

3.5 "Burgulence"

"Burgulence" conflates "Burgers" with "turbulence" and means "the study of random solutions to the Burgers equation," this is how a review defines this approach towards the turbulence problem (Frisch and Bec 2001). The attempt was launched by Burgers in 1939 with "Mathematical examples illustrating relations occurring in the theory of turbulent fluid motion" (Burgers 1939), followed by sequels that exposed the pertinence to statistical theories of turbulence (Burgers 1940a, b). What became known as "Burgers equation" is a set of partial differential equations that model various flow configurations. Plane channel flow, for example, was modelled by

---

[8]Prandtl to Ludwig Hopf, 15 October 1938. AMPG, Abt. III, Rep. 61, Nr. 2153. Translation ME.

$$b\frac{dU}{dt} = P - \frac{\nu U}{b} - \frac{1}{b}\int_0^b dy\, u^2$$

$$\frac{\partial u}{\partial t} = Uu + \nu\frac{\partial^2 u}{\partial y^2} - 2u\frac{\partial u}{\partial y}$$

where U models a mean speed in x-direction limited by plane walls at $y = 0$ and $y = b$, $u(y)$ plays the role of a "secondary motion" which vanishes at the boundaries at $y = 0$ and $y = b$, and P and ν are constants modelling an external force and friction, respectively. (For a detailed account on this model see Burgers (1939, pp. 14–18), and Burgers (1948, pp. 172–177); the latter is a more accessible account in which Burgers reviewed his models from 1939 and 1940).

Burgers's models represented a simplification of the three-dimensional Navier-Stokes equations in the hope to preserve the principal features of turbulence. Burgers was conscious that the model equations could not account for "those properties of the turbulent field, which are considered e.g. in the theories of Taylor, von Kármán and others on isotropic turbulence and related problems." Yet he hoped they "may lead to certain typical results to which a statistical theory of such systems should conform" (Burgers 1939, pp. 1–2). The channel flow model, for example, gave rise to a "laminar solution" ($U = \frac{Pb}{\nu}$, $u = 0$) and "stationary turbulent solutions" for which he derived special equations (Burgers 1939, p. 18).

But this approach evoked little enthusiasm among the pioneers of the statistical theory of turbulence. Although Burgers entertained a lively correspondence with Taylor in the 1930s on the statistical theory, there was no discussion on Burgers's mathematical model equations. In this regard "Taylor and Burgers were not on the same wavelength (despite the mutual fondness)," an expert on Taylor's research on turbulence concluded from their mutual exchange of letters (Sreenivasan 2011, p. 157). The same may be true for Prandtl and Kármán. Burgers could not offer practical results referring to wind tunnel turbulence or other applications—in contrast to the contributions of Prandtl, Taylor and Kármán. Another reason for the contemporary neglect of Burgers's 1939 treatise may be the outbreak of the Second World War in the same year.

Nevertheless, even during the war, Burgers's contribution arouse some interest. When Sommerfeld announced in January 1945 in a letter to Prandtl his forthcoming textbook on *The mechanics of continuous media*, he remarked that he dealt with "the ominous turbulence" in several sub-chapters, and that he regarded Burgers's mathematical model as a remarkable contribution.[9] Already in the introduction to the chapter on turbulence Sommerfeld posed what he regarded as the basic question about the turbulence problem: Are the Navier-Stokes equations sufficient to explain the observed turbulence phenomena? "Even such outstanding experts like Th. von Kármán and G. I. Taylor tend in some of their work to the view that, like in the theory of gases, one can cope with turbulence only by statistical methods." But Sommerfeld answered this question in the affirmative. "The example that we owe to Burgers can support us in the view that turbulence belongs to classical hydrodynamics, even

---

[9]Sommerfeld to Prandtl, 31 January 1945. MPGA, Abt. III, Rep. 61, Nr. 1538.

though to a specialty that is hard to access mathematically."[10] In a sub-chapter titled "On the basic problem of turbulence" he analysed the "extraordinarily simplified simile (Gleichnis) on the processes of turbulence" in the form of

$$\frac{dU}{dt} = P - \nu U - u^2$$
$$\frac{\partial u}{\partial t} = Uu - \nu u.$$

These equations were "nonlinear like the turbulence equations" (due to u^2 in the first equation and Uu in the second) and they accounted for dissipation (due to νU in the first and νu in the second equation). The analysis resulted in a "laminar" solution ($U = P/\nu, u = 0$) for $P < \nu^2$ and a "turbulent" solution for $P > \nu^2$. The lesson from Burgers's "method of 'mathematical models'" was clear: "The nonlinear character of the equations gives rise, here like in turbulence, to the possibility of entirely different modes of motion."[11]

In 1944, when Sommerfeld concluded his chapter on turbulence in this way, he may have been alone with his praise of Burgers's mathematical similes. But the surge of interest in the Burgers equation after the war (see Sect. 5.1) and its further development into "Burgulence" justify the attention which Sommerfeld paid to this effort in the history of turbulence problem.

---

[10] Sommerfeld (1945, pp. 261). Translation ME.
[11] Sommerfeld (1945, pp. 269–272). Translation ME.

Chapter 4
Turbulence in WW II

Abstract Research on turbulence became part of secret war projects. The development of "laminar wings" with minimal skin friction involved investigations about the transition to turbulence in the boundary layer. Fully developed turbulence was investigated because of its impact on heat transfer and other practical applications. Nevertheless, progress was also achieved in more fundamental areas with little or no relation to wartime projects. In 1941 Kolmogorov published a theory which predicted for the statistics of the smallest scales of turbulence at very high Reynolds numbers a universal behaviour (K41-theory). Prandtl presented an energy equation for fully developed turbulence and arrived (in unpublished notes) at the same microscale of turbulence as Kolmogorov.

International exchange, such as in 1938 at the symposium on turbulence during the Fifth International Congress for Applied Mechanics, came to an end with the outbreak of the Second World War in September 1939. Applied mechanics, and fluid mechanics in particular, was pertinent to a host of war technologies, from submarines to airplanes, bombs and rockets. In many cases the development of advanced weapons involved research on turbulence, such as the transition to turbulence in boundary layers or turbulent heat transfer.

Most of this research was of an applied character which benefited from established pre-war knowledge on turbulence.[1] However, in some cases new applications provoked research that touched on old riddles like the onset of turbulence. In a few cases, basic research on turbulence was pursued despite the war—almost like a private escape into more challenging riddles.

---

[1] See Eckert (2018) for a discussion of turbulence research in the interwar period with regard to the "applied" versus "fundamental" dichotomy.

© The Author(s), under exclusive license to Springer Nature Switzerland AG 2019　　39
M. Eckert, *The Turbulence Problem*,
SpringerBriefs in History of Science and Technology,
https://doi.org/10.1007/978-3-030-31863-5_4

4.1 Kolmogorov's Statistical Theory

A breakthrough of the latter kind occurred in Russia in 1941. Andrey Nikolaevich Kolmogorov, a mathematician and member of the Moscow Academy of Sciences, published in this year a theory about turbulence at very high Reynolds numbers that became labeled later as "K41-theory". Still in 2003, at the centenary of Kolmogorov's birth, this theory was considered "among the most important contributions in the long-standing history of the theory of turbulence" (Livi and Vulpiani 2003, p. VIII). At first sight one might be tempted to associate this theory with Kolmogorov's scientific war-work. When the German troops invaded Russia in 1941, Kolmogorov applied his skills as a statistician: he suggested a stochastic distribution of barrage balloons over Moscow as a protection against bombing (Chentsov 1990, p. 987). But a closer look at the prehistory of K41 shows that Kolmogorov's theory of turbulence had other roots that were unrelated to his war effort.

Kolmogorov was a mathematician with a wide range of interests. Turbulence was just one of them. It may have been kindled in 1939 when Kolmogorov became a member of the academy and made secretary of the section of Physics and Mathematics of a new Institute of Theoretical Geophysics. In the same year, Kolmogorov asked his student Mikhail D. Millionshchikov to investigate the decay of turbulence. In 1940 he involved another student, Alexander Mikhailovich Obukhov, who focused on the distribution of energy in fully developed turbulence. From the collaboration with these students, and the study of previous statistical theories, Kolmogorov developed a scheme how energy and momentum was transferred to smaller and smaller scales. At the smallest scale viscosity would take over and transform kinetic energy into heat. With few assumptions (local homogeneity; constant mean energy dissipation rate; statistics of velocity differences independent on viscosity) he derived the scaling laws for turbulence that now bear his name.[2]

It is interesting to note that Kolmogorov did not address turbulence as the outstanding problem to which so many of his predecessors had contributed frustrated efforts. In his academy communications he did not pretend to solve a great riddle. Even in retrospect, when his theory was celebrated as a major breakthrough, Kolmogorov described it as a "purely phenomenological" theory. As one of his pupils, Yakov G. Sinai, recalled (Livi and Vulpiani 2003, p. V):

> When Kolmogorov was close to eighty I asked him about the history of his discoveries of the scaling laws. He gave me a very astonishing answer by saying that for half a year he studied the results of concrete measurements. In the late Sixties Kolmogorov undertook a trip on board a scientific ship participating in the experiments on oceanic turbulence. Kolmogorov was never seriously interested in the problem of existence and uniqueness of solutions of the Navier-Stokes system. He also considered his theory of turbulence as

---

[2]For historical accounts see Yaglom (1994) and Falkovich (2011); Kolmogorov's academy papers from 1941 have been translated into English and published in 1961 as "classic papers on statistical theory" (Friedlander and Topper 1961); two of his 1941 papers were included again in 1991 in a special volume of the *Proceedings of the Royal Society* at the occasion of the fiftieth birthday of K41 (Kolmogorov 1991b, a); a textbook introduction is provided by Frisch (1995).

purely phenomenological and never believed that it would eventually have a mathematical framework.

K41, therefore, was not the discovery of a lone mathematician devoid of empirical underpinning. Nor was it a side-product of Kolmogorov's war work as a statistician. The measurements to which Kolmogorov alluded (Kolmogorov 1991a, p. 17) were those on wind tunnel turbulence of Dryden's group at the National Bureau of Standards from the mid-1930s (Dryden et al. 1936).

Outside the USSR Kolmogorov's papers became known only after the war (see Sect. 5.1). Among the circle of his pupils and fellow academicians, however, they provoked further debates. In January 1942 Kolmogorov delivered a report on "Equations of turbulent motion of an incompressible fluid" at a seminar in Kazan, a town on the Volga where the Academy had been evacuated. Among the participants were also the physicists Pyotr Leonidovich Kapitsa and Lev Davidovich Landau. There is little historical evidence about the Kazan debates, except a brief resumé of Kolmogorov's presentation. Kolmogorov's equations were a set of differential equations for the computation of the mean fluctuating turbulent motion. "The solution of these equations presents great difficulties," Kolmogorov exercised caution. Landau criticized that Kolmogorov's equations did not account for the conservation of vorticity in limited regions (Spalding 1991, pp. 215–216). There has been some speculation about Landau's reaction to Kolmogorov's theory. With regard to the treatment of turbulence in the famous Landau-Lifshitz textbook on *Fluid Mechanics* it has been argued that Landau doubted the possibility of a universal solution for fully developed turbulence (Frisch 1995, Sect. 6.4).

Whatever arguments may have been exchanged at Kazan between Landau, Kapitza and Kolmogorov, they called for further investigations about the problem of fully developed turbulence. In 1944 Landau introduced a note "On the problem of turbulence" in the Academy proceedings with the remark that despite extensive studies on turbulent motions "the very essence of this phenomenon is still lacking sufficient clearness." He suggested that "the problem may appear in a new light if the process of initiation of turbulence is examined thoroughly" (Landau 1944, p. 387). He perceived fully developed turbulence as an accumulation of quasi-periodic motions. "In the course of a further increase of the Reynolds number more and more new periods appear in succession, and the motion assumes an involved character typical of a developed turbulence" (Landau 1944, p. 390).

4.2 Laminar Wings

By the same time a thorough investigation of periodic motions in the boundary layer on a flat plate was pursued in the USA as part of war research. The quest for wings with minimal drag for pursuit planes involved research on turbulence because the profile drag of a wing depends of the length of the laminar and turbulent air flow in the boundary layer. A so called "laminar wing" is shaped in such a way that the transition to turbulence in the boundary layer is delayed as far as possible away from the leading edge of the wing to the rear. Since 1940 Dryden's Aerodynamics Division at the National Bureau of Standards pursued projects under NACA-contract in order to explore "Transition Phenomena at Low Turbulence" and "Methods of Reducing Wind Tunnel Turbulence," because the turbulence of the air stream in the wind tunnel affects the measurements of laminar wing profiles. The goal was to determine "the effect of small intensities of wind tunnel turbulence on the position of transition from laminar to turbulent boundary-layer flow." The effect should be measured at first on a flat aluminium plate exposed to the air stream of the low-turbulence tunnel. Because other measurements had indicated that even at very low levels of wind tunnel turbulence a transition could be evoked, "emphasis will be laid on reducing the turbulence as much as possible and investigating transition at the lowest values obtainable."[3]

The prevailing view was "that turbulence acts as a boundary layer disturber (Taylor 1939, p. 307)" (see Sect. 3.4). Reduction of wind-tunnel turbulence, therefore, would remove the "disturber" and allow for a precise determination of the position of the laminar–turbulent transition in the boundary layer. But in May 1941 Dryden's group reported a rather surprising phenomenon: "Under conditions of low stream turbulence, transition is found to result from growth in amplitude of a fairly regular oscillation of the boundary layer." Instead of the external "disturber" the initiation of turbulence was due to oscillations within the boundary layer. "The oscillations have some of the characteristics of those predicted by the Tollmien-Schlichting theory which have not heretofore been experimentally observed. The frequencies are in fact in line with those predicted. Likewise the speed fluctuation in the outer portion of the boundary layer is in opposite phase to that near the wall as predicted by the theory." What later would be called "Tollmien-Schlichting waves" was designated in this report as "a kind of boundary layer 'flutter'," alluding to the phenomenon of wing flutter.[4]

Henceforth the "natural oscillations of the boundary layer which have been studied by Tollmien and Schlichting" became subject of further experimental investigation. The goal of laminar wings turned the "Investigation of Laminar Boundary-Layer Oscillation," as the research contract for the fiscal year 1942 was titled, into a war research of practical interest.[5] Without the prospect of application "to current problems relative to military aircraft both to wing development and to ducting problems as well," as the NACA headquarters proclaimed in February 1943, these investigations would not have been funded as war research (Roland 1985, vol. 2, p. 549). The experimental results were communicated in April 1943, together with a review of the Tollmien-Schlichting theory, as a NACA Advance Confidential Report and published without restrictions after the war as NACA Report No. 909 (Schubauer

---

[3]Lewis to Warner, 6 May 1940. Hugh L. Dryden papers, ms. 147, Special Collections, Milton S. Eisenhower Library, The Johns Hopkins University (Dryden papers), subject files, misc. correspondence, box 62.

[4]Report to NACA, 10 May 1941. Ibid.

[5]Proposal to NACA for the Fiscal Year 1942. Ibid.

and Skramstad 1943). Dryden's group at the National Bureau of Standards was not the only NACA-contractor for boundary layer research in low-turbulence wind tunnels. At the GALCIT, Kármán entrusted in 1942 his student, Hans Liepmann, with similar investigations—and Liepmann, too, found the oscillations predicted by the Tollmien-Schlichting theory, both on flat and curved plates (Liepmann 1943).

In Germany laminar wing research was solicited in 1940 by the Air Ministry with a prize competition: "For the further improvement of aircraft performance it is necessary to reduce the drag of all surfaces exposed to the airstream." The competition was tendered by the Lilienthal Society, an organization under the umbrella of the Air Ministry. "According to the results of research on friction layers, successes may be expected if the boundary layer is kept laminar as far as possible downstream; the transition point beyond which the boundary layer is turbulent should be moved away as far as possible from the stagnation point at the leading edge." The prize committee, chaired by Tollmien, particularly called for investigations aimed at predicting the transition from laminar to turbulent boundary layer flow in terms of its "dependence on the shape of the body or profile of the wing". The Lilienthal Society published the awarded research papers in a secret wartime report.[6] The first prize was awarded to Schlichting and his assistant, Albert Ulrich. Based on the Tollmien-Schlichting theory for a flat plate, they developed a numerical procedure for arbitrary wing profiles.

In April 1941, Schlichting reviewed the contemporary knowledge about the laminar–turbulent transition and its pertinence for laminar airfoils in a seminar at the Luftfahrtforschungsanstalt Hermann Göring, a secret aeronautical research facility in Völkenrode near Braunschweig. He described the laminar wing problem as part and parcel of the turbulence problem concerned with the onset of turbulence—and therefore accessible to theoretical solutions[7]:

> Summarizing I would like to assert that the problem of the onset of turbulence may be regarded from the perspective of theory as principally solved. The method of small oscillations yields satisfactory results if one takes into account in the right manner the physically important quantities, the curvature of the profile and the viscosity. [...] Solved in principle is also the practically important problem to compute the point of transition for wing profiles [...]
> *It is possible today to compute for an arbitrary wing profile the dependency of the transition point on the Reynolds number $U_0 t/\nu$, the lift coefficient c_a and therefore also on the shape of the wing section.*

In October 1941, an expert committee of the Lilienthal Society convened a conference on "Boundary-Layer Problems" where Schlichting and his collaborators presented a systematic study about the influence of the shape of a profile on the point where the laminar boundary becomes turbulent (Lilienthal-Gesellschaft 1941). In the following winter semester 1941/42 Schlichting made "Boundary Layer Theory" the subject of a lecture series at the Luftfahrtforschungsanstalt Hermann Göring. It became the blueprint of his textbook with the same title (Schlichting 1942, 1949a,b, 1951). Although Schlichting focused on the theoretical analysis,

---

[6]Lilienthal-Gesellschaft (1940). Translation ME.
[7]Schlichting (1941, p. 30). Translation ME.

the research involved also experimental investigations. In December 1942, Schlichting's assistant presented in a war report the results of aerodynamic measurements of a number of laminar profiles performed in wind tunnels of the Technische Hochschule Braunschweig and the Luftfahrtforschungsanstalt Hermann Göring in Völkenrode (Bußmann 1942). Other investigations were reported in January 1943 from the Deutsche Versuchsanstalt für Luftfahrt in Berlin-Adlershof (Doetsch 1943). Laminar wings of captured pursuit planes, such as the "P-51 Mustang," were subjected to wind tunnel tests in various facilities (Bußmann 1943; Breford and Möller 1943; Riegels 1943).

Despite considerable research it remained an open question whether laminar wings displayed under actual flight conditions the low drag expected from theoretical and experimental investigations. Besides a low degree of turbulence the comparison with real flight conditions required extremely high Reynolds numbers which apparently could not be attained in the available wind tunnels. Although a special low turbulence wind tunnel at the AVA became operational in 1943, it was apparently not used for the development of laminar wings. "This influence (of the degree of turbulence in the wind tunnel) could only be clarified by measuring the same laminar profiles in the wind tunnels and for comparison in free flight or in a wind stream of very low turbulence," a postwar report reviewed the German war research on laminar wings. "Such measurements have not been conducted in Germany heretofore. For this reason we cannot answer the crucial question whether practical laminar profiles can be realized at the Reynolds numbers of today's airplanes."[8]

Whether or not laminar wings would become practical assets, they contributed to make turbulence a subject of war research—and influenced the perception of the "turbulence problem." Schlichting distinguished in his wartime lectures "the two main problems" in the theory of turbulence as "The flow laws of the developed turbulent flow" and the "Origin of turbulence." The latter amounted to "a stability investigation, made to clarify theoretically the laminar-turbulent transitlon." Despite Schlichting's own contributions to this research, however, he assigned greater priority to the former. His focus on fully developed turbulence was less motivated by fundamental considerations than practical applications. The main subject of his lectures concerned fully developed turbulence in the boundary layer. "The problem is to calculate the local distribution of the time average of the velocity components, and thus to gain further information concerning, for instance, the friction drag" (Schlichting 1949b, pp. 4–5). His audience of aeronautical engineers at the Luftfahrtforschungsanstalt Hermann Göring hardly needed further remarks about the applications to which Schlichting alluded.

---

[8]Holstein (1947, p. 13). Translation ME.

4.3   Turbulence Problems in Miscellaneous War Applications

The variety of applications makes it difficult to categorize research on turbulence during the war. In a postwar review Prandtl distinguished between "Wall-bounded turbulence", "Free turbulence" and "Individual problems" (Prandtl 1953, p. 55). Another report of German wartime research on turbulence lists these topics: "Stability theory of laminar flows", "Turbulent friction layers", "Free turbulence", "A series of fundamental investigations" and "Geophysical applications" (Görtler 1948, p. 75). The *Monographien über Fortschritte der deutschen Luftfahrtforschung (seit 1939)*, solicited by the British military government and edited by the director of the AVA in Göttingen, Albert Betz, reviewed research on turbulence in a volume on "Boundary layers" under the headline "Turbulent friction layers".

War research on turbulence, therefore, could appear under different umbrellas. As Kármán had observed few years before the war, "if we meet a practical question in aerodynamic design which we are unable to answer, the reason that we are unable to give a definitive answer is almost certainly that it involves turbulence" (von Kármán 1937b, p. 1141). "Turbulent friction layers," for example, were not only a concern for skin friction but also for the heat transfer between a hot solid body and the surrounding flow. In December 1940, Prandtl's close collaborator on turbulence research, Hans Reichardt, had published "a general theory of turbulent heat transfer, from which the technically important special cases may be derived easily" (Reichardt 1940, p. 297). Schlichting dedicated one of his Völkenrode lectures in 1942 to the "Connection Between Exchange of Momentum, Heat and Material" (Schlichting 1949b, pp. 66–68). Karl Wieghardt, another expert on turbulence in Prandtl's KWI throughout the war years, accounted for "Turbulent flow" in another postwar review on "Heat transfer" (Wieghardt 1948).

A good deal of research described in these reviews was concerned with fully developed turbulence in free jets and in the turbulent boundary layer along the surface of solid bodies. The latter became a particular subject of war research. At Prandtl's KWI these inverstigations were pursued in a special "roughness wind tunnel" ("Rauhigkeitskanal") where the turbulent boundary layer over selected surfaces could be measured at prescribed pressure distributions (Schultz-Grunow 1940). In 1940 it was expanded and adjusted for the use of war contracts. As Prandtl informed the Air Ministry in May 1940, this wind tunnel was "now permanently in operation, mainly for investigations by order of aircraft companies."[9] A proposal for research funds in March 1942 specified the "blowing of air into the friction layer" as one of the major uses of the "roughness wind tunnel"[10]:

> Air enters accidentally from the inner of an aircraft into the friction layer of fuselage and wings at leakages and joint fissures, furthermore at the cladding of engines etc. Industry is very much interested in a systematic investigation of such disturbances of flow. On the other side one attempts to exert a favourable influence on the friction layer of wings by

---

[9]Prandtl to RLM, 25 May 1940. AMPG, Abt. I, Rep. 44, Nr. 45.

[10]Attachment to the proposal of 21 May 1942. AMPG, Abt. I, Rep. 44, Nr. 46. Translation ME.

blowing air into it with special arrangements of flaps, so that the lift is increased and the drag reduced, respectively. The investigations which have become known so far refer only to special arrangements from which no general conclusions may be drawn. Therefore, the whole range of these processes should be clarified by experiments which will be pursued in the roughness wind tunnel of the KWI. It is planned to perform drag measurements up to Reynolds numbers of about 10^7 by measuring the momentum at prescribed pressure distributions inside the measuring chamber (constant pressure, increasing and decreasing pressure).

These and other investigations in the roughness wind tunnel required a sophisticated experimental technique. "Development of measurement apparatus for turbulence" was a regular item on Prandtl's list of research proposals to the Air Ministry throughout the war. Further war contracts which involved experiments in the roughness wind tunnel came from the Navy.[11] A particular investigation concerned the skin friction of rubber layers used as a camouflage for submarines against acoustic detection (Wieghardt 1944). Another war contract issued by the Naval Observatory in Greifswald was dedicated to the problem "how a gas that is blown out uniformly close to the ground spreads over a plane surface". Because Prandtl's turbulence expert, Karl Wieghardt, could hardly experiment with warfare agents, he resorted to the analogy between the turbulent mixing of gas and the turbulent spreading of heat in a current of air. Thus the investigation was performed with thermocouples for measuring the distribution of temperature in the wake of a heater coil placed in the bottom of the roughness wind tunnel.[12]

War related research on turbulence was pursued also in other countries. In the U.S.A. the quest of low-drag airfoils resulted in the construction of a giant "two-dimensional low turbulence pressure tunnel" at the NACA Langley Laboratory. Although it was dedicated to the measurement of the characteristics of low-drag airfoil sections at high Reynolds numbers and low degrees of turbulence (Tetervin 1943; Tucker and Wallace 1944), the operation of this tunnel involved detailed investigations of the behaviour of turbulent boundary layers (Doenhoff and Tetervin 1943). At the request of NACA, other studies of turbulent flows were performed by Kármán and his collaborators at the GALCIT in Pasadena. Besides Liepmann's experiments on the transition to turbulence which confirmed the Tollmien-Schlichting theory (see Sect. 4.2), Stanley Corrsin began his career in turbulence research in this context with investigations of turbulent jets (Corrsin 1943, 1944).

Although there was little incentive to focus on fundamental problems, at least in one case the results of classified war research suggested a revision of former theoretical convictions: When the disputed Tollmien-Schlichting theory was twice corroborated by Dryden's and Kármán's groups at the National Bureau of Standards and at GALCIT, Kármán asked a Chinese student, Chia-Chiao Lin, to review the Orr-Sommerfeld approach. Lin was at that time not yet a naturalized US citizen and thus not cleared for confidential war research. Therefore, unlike Liepmann's experi-

---

[11]Reports by Prandtl on war contracts, 1942–1944. AMPG, Abt. I, Rep. 44, Nr. 46, 48, 52 and 70. See also Epple (2002).

[12]K. Wieghardt: Über Ausbreitungsvorgänge in turbulenten Reibungsschichten. Geheimbericht für das Marineobservatorium Greifswald, 1. September 1944. APMG, Abt. III, Rep. 76B, Kasten 2. For more detail, see Schmaltz (2005, pp. 326–356).

mental investigation, his revision of the stability theory was not officially performed under a war research contract. Nevertheless, Kármán sent Dryden a diagram "showing Lin's curve (present calculation) for the stability limit in the boundary layer compared to Schlichting and the experimental evidence." Lin's curve supported the conviction that both groups had indeed observed the disputed Tollmien-Schlichting waves. "Unfortunately because of the confidential character of yours and Liepmann's work this diagram cannot be published," Kármán regretted in a letter to Dryden.[13] Lin's work became published without reference to the experimental war research as a doctoral dissertation (Lin 1944). Thus the turbulence problem as perceived in the wake of the Orr-Sommerfeld approach found a late answer. This case illustrates how closely fundamental science could be related to applied war research.

4.4 Fundamental Wartime Research on Turbulence

Contrasting "fundamental" and "pure" from "applied" and "goal-oriented" research is often motivated by ideology. The former appears as innocent in contrast to the latter which may be tainted by dubious interests. Yet it is clear that certain investigations on turbulence were closer to wartime application than others, at least on the short term. Kolmogorov's statistical theory of turbulence and Prandtl's measurements in the roughness wind tunnel both involved fundamental questions of fully developed turbulent flow, but they were quite different in terms of wartime application. Prandtl introduced his review on German turbulence research between 1939 and 1945 with the remark, that "various more subtle investigations had to be abandoned for war-related reasons" in favour of "special problems of mainly practical importance" ((Prandtl, 1953), p. 55). Thus Prandtl made it appear as if he was forced to abandon more fundamental research on turbulence. In the beginning of the war, however, he had deliberately refrained from research which he regarded as unimportant for the war. He cancelled investigations on turbulent fluctuations from his agenda, for which the Air Ministry had just funded a new wind tunnel at his KWI. "The wind tunnel should rather be reserved for war-important work at the AVA," Prandtl diverted its use.[14] Although the investigation of turbulent fluctuations had reached center stage since the mid-1930s, Prandtl regarded it less important for war applications.

With regard to theoretical study, however, Prandtl did not completely abstain from pursuing his pre-war agenda. In autumn 1944 he elaborated some of the ideas which he had presented in 1938 at the turbulence symposium. His focus was on the energy per unit volume contained in the turbulent fluctuations in fully developed turbulence. This approach would later become categorised as "One-equation turbulence model" (see Sect. 6.4). On 4 January 1945 Prandtl presented preliminary results at an internal colloquium to the collaborators of his institute; three weeks later he submitted to

---

[13] Kármán to Dryden, 26 February 1944. Dryden papers, folder: 1935–1959. Misc. Dryden–Von Kármán Correspondence.

[14] RLM-Programm 1940. AMPG, Abt. I, Rep. 44, Nr. 45. Translation ME.

the Göttingen Academy of Science a treatise "Über ein neues Formelsystem für die ausgebildete Turbulenz" ("On a new system of equations for fully developed turbulence") for publication in the Academy's proceedings (Prandtl and Wieghardt 1945).

Prandtl's unpublished notes show that he aimed in 1944/1945 at a general theory of developed turbulence. In this venture he also arrived at the result for the downgrading of turbulent motions to "Kolmogorov's micro-scale" (as they were later named):

$$\lambda = (\frac{\nu^3}{\varepsilon})^{1/4} \quad u' = (\nu\varepsilon)^{1/4}$$

where λ is the Kolmogorov length, ν the kinematic viscosity, ε the turbulent kinetic energy per time and unit volume that passes from large to small scales, and u' the velocity fluctuation at the smallest scale. Prandtl found this result so notable that he added the remark: "Checked it multiple times! But only equilibrium of turbulence."[15]

Prandtl's focus on basic problems of turbulence during the war does not contradict his intent to divert his research on applied problems in the beginning of the war. "Various recent observations, for example those about the turbulent friction layer on plates [...] call for a new computational approach." In this way he revealed in October 1944 that the measurements in the roughness wind tunnel motivated his new theoretical investigations on fully developed turbulence.[16] At the turbulence colloquium in 1938 he had exposed the discrepancy between theoretical approaches for "wall turbulence", "free turbulence" and "isotropic turbulence" as a particular problem. Now he focused on the kinetic energy of the turbulent fluctuations because he hoped to arrive in this way at a uniform mathematical framework—for which the wartime measurements provided some empirical underpinning. His energy equation was supposed to account for all sorts of fully developed turbulence, only the constants therein had to be adjusted respectively. "The most important practical task will be to derive from appropriate experiments preliminary numerical values for the three essential constants," Wieghardt introduced an appendix to Prandtl's theory. He derived these values from his own experiments in the roughness wind tunnel and other recent experimental measurements for grid-generated turbulence and channel flow, respectively, and thus put Prandtl's theory on solid empirical ground (Prandtl and Wieghardt 1945, pp. 14–19).

The "fundamental" versus the "applied" or "war-related" research on turbulence was therefore not as far apart from another as the publication in the proceedings of the Göttingen Academy of Science versus secret war reports suggests. Both contributed to a more pronounced perception of the turbulence problem as a riddle whose ramifications in a host of quite different phenomena did not preclude the quest for a uniform theoretical solution.

---

[15]Page 45 of Prandtl's unpublished notes on "Ausbreitungstheorie der Turbulenz", GOAR 3727. This page is headlined "Einbeziehung der Zähigkeit" and dated "29. 1. 45". In another manuscript he analysed the role of viscosity in the mechanism of developed turbulence more closely, GOAR 3712. For more details on Prandtl's unpublished investigation see Bodenschatz and Eckert (2011, Sect. 2.10).

[16]Page 1 of Prandtl's unpublished notes on "Ausbreitungstheorie der Turbulenz", GOAR 3727. Translation ME.

Chapter 5
Expectations and Hopes: 1945–1961

Abstract The revelation of wartime achievements raised hopes that the turbulence problem is close to a solution. The turbulence problem of the 1920s—the instability of laminar flow—had been solved for the flat plate (Blasius flow) and theoretically also for plane Poiseuille flow. The statistical theory of fully developed turbulence as elaborated by Kolmogorov and others had achieved remarkable success. In the 1950s, the first textbooks portrayed turbulence as a subject of research in its own right. In 1961, an international conference on turbulence signalled its impact mainly on atmospheric and geophysical applications.

As soon as Nazi-Germany was defeated specialized allied intelligence units began to survey the achievements made by German scientists and engineers in their wartime research—including turbulence. The Göttingen Aerodynamics Research Institute and the Kaiser-Wilhelm-Institute for Fluid Dynamics, the still widely admired home of Prandtl and his school, was frequently visited by one or another group of experts, among them Prandtl's close colleagues Theodore von Kármán and Hugh Dryden (Fig. 5.1). Here they learned from personal interrogations about Prandtl's wartime research (Eckert 2017, Sect. 9.1).

The broader scientific community learned about wartime research on turbulence for the first time at the Sixth International Congress for Applied Mechanics, held in September 1946 in Paris. Like in the preceding congress before the war, a special symposium was dedicated to turbulence.

5.1 A "Remarkable Series of Coincidences"

For most participants at the Paris congress the revelation of war-time achievements in turbulence research came as a complete surprise. Hugh Dryden aroused considerable interest with his conference paper on "Some Recent Contributions to the Study of Transition and Turbulent Boundary Layers" in which he disclosed the discovery of Tollmien-Schlichting waves in his laboratory at the National Bureau of Standards.

© The Author(s), under exclusive license to Springer Nature Switzerland AG 2019   49
M. Eckert, *The Turbulence Problem*,
SpringerBriefs in History of Science and Technology,
https://doi.org/10.1007/978-3-030-31863-5_5

Fig. 5.1 In May 1945 Hugh Dryden (left), Theodore von Kármán (with cigar) and others visited German aeronautic research facilities. At this opportunity they also learned about Prandtl's recent efforts concerning the turbulence problem

G. I. Taylor recalled in a short article in *Nature* the controversial pre-war debates about the transition to turbulence. Now this riddle seemed resolved: "Dryden has now shown that if the air stream in a wind tunnel is sufficiently free from turbulence, an instability of the boundary layer which had been predicted mathematically does, in fact, appear at the calculated wind-speeds, and that it has the calculated frequency and wave-length. These unstable waves are masked by larger effects in wind tunnels not specially designed to be free of turbulence" (Taylor 1946, p. 775).

Another revelation concerned "some recent theories regarding the correlation coefficient in isotropic turbulence at high Reynolds numbers," as Dryden reported in *Science*. "Three scientists, Kolmogoroff in the U. S. S. R., Onsager in the United States, and C. F. Weizsäcker in collaboration with W. Heisenberg in Germany, obtained the same result" (Dryden 1947, p. 169). Weizsäcker's and Heisenberg's theories were unpublished. Onsager had published only a short abstract. No one of them attended the Paris congress. It was George Batchelor who informed the congress about their theories. He was the first outside Russia who studied Kolmogorov's papers from 1941, and he had learned about Weizsäcker's and Heisenberg's work from oral conversations with his mentor, G. I. Taylor. Now he regarded it as his mission to disseminate the news at the Paris congress. "The three theories have certain elements in common, sometimes explicit and in some cases implied," Batchelor summarized his impression about this "remarkable series of coincidences". The "physical picture of turbulence" was "roughly identical", but "the mathematical formulations vary considerably" (Batchelor 1946, p. 883).

The almost simultaneous emergence of "roughly identical" theories of turbulence has become subject of special inquiry (Battimelli and Vulpiani 1982). The different motivations and approaches underlying these theories illustrate once more the wide range of concerns and contexts from which fundamental research on turbulence could emerge. Onsager, for example, aimed at a theory of point-vortex dynamics. "I received a letter and a kind of manuscript from a certain Mr. Lars Onsager," Kármán wrote in August 1945 to his doctoral student C. C. Lin who had pursued a similar approach. "I find his letter somewhat 'screwy' so I would be glad to have your opinion". Lin regarded Onsager's approach as "not fully developed, if there is something to be found behind his idea," although he admitted that Onsager had achieved "many good things in his line (statistical mechanics, thermodynamics, etc.)" (quoted in (Eyink and Sreenivasan 2006, p. 116)). Thus Onsager's "screwy" effort was largely ignored by his contemporaries. It was found only later that his private notes from the 1940s contain significant results. With regard to his publications, however, Onsager's excursion into turbulence remained solitary. He became famous for other achievements. In 1968 he was awarded the Nobel prize in chemistry "for the discovery of the reciprocal relations bearing his name, which are fundamental for the thermodynamics of irreversible processes."[1]

Weizsäcker's and Heisenberg's work on turbulence was kindled in summer 1945 when the Allies detained them together with eight other German nuclear physicists in order to learn more about their knowledge of atom bombs. The detainees spent some of their time with self-organized colloquia. Weizsäcker put turbulence on the agenda because of his interest in the origin of the planetary system. He supposed that turbulence plays a crucial role within the circumsolar gas disk from which the sun and the planets would form. When Weizsäcker mentioned in one of these colloquia Burgers's model equations "a very nice idea cropped up in the discussion", one of the detainees noted in his diary, "namely: the loss of energy per unit of time is the same for each Fourier term of the turbulent velocity. Can one put into effect the theory of turbulence by assuming this theorem as valid in the three-dimensional case? What consequences result from this assumption?"[2] At Farm Hall, the country house where they spent the final months of their detention, Heisenberg and Weizsäcker developed the theory in more detail. According to Batchelor, they also discussed about their work with G. I. Taylor (Batchelor 1996, pp. 171–172). After they were released to Germany in 1946, Weizsäcker and Heisenberg submitted their theories to the *Zeitschrift für Physik*, where they appeared with a time lag of almost two years in 1948 as consecutive articles (Weizsäcker 1948; Heisenberg 1948b).

If Batchelor had known about Prandtl's manuscript in which the dissipation length scale was independently derived during the last months of the war (see Sect. 4.4), he would certainly have added it to the "remarkable series of coincidences". Prandtl mentioned it briefly at the end of a review of German war-work on turbulence (Prandtl 1953, p. 77), but he did not care to publish it—perhaps because he regarded the

---

[1] https://www.nobelprize.org/prizes/chemistry/1968/onsager/facts/ (25 October 2019). For Onsager's short excursion in turbulence see Onsager (1945).
[2] Bagge et al. (1957, p. 49). Translation ME.

subject by then already elaborated in exhaustive detail by Weizsäcker and Heisenberg. Both were in close contact with Prandtl after they had come to Göttingen in 1946 with the mission to re-construct physics in Germany under the authority of the British military government.

5.2   The Turbulence Problem Ca. 1950

The fact that *Nature* and *Science* reported about the Paris congress in 1946 mirrors a growing awareness among scientists for the recent advances in fluid mechanics in general, and turbulence in particular. Shortly after the congress Burgers, Kármán and like-minded advocates of international science established the International Union of Theoretical and Applied Mechanics (IUTAM) as a new umbrella for the international mechanics congresses. Among IUTAM's first activities was the organization of the Seventh International Mechanics Congress scheduled to take place in September 1948 in London—two years earlier than according to the accustomed four-year interval in order to avoid interference with the International Congress of Mathematicians scheduled for 1950. Batchelor presented the general lecture on "Recent developments in turbulence research" (Batchelor 1948). Other speakers who focused on one or another aspect of turbulent flows were Hugh Dryden, Alexandre Favre, Francois Naftali Frenkiel, C. C. Lin and J. O. Hinze.

Batchelor had already before the London congress published more details of Kolmogorov's theory (Batchelor 1947). The papers of Weizsäcker and Heisenberg, too, had been circulated among experts before they appeared in print. Kármán referred to them as "unpublished" when he dedicated a lecture in June 1948 on "Progress in the Statistical Theory of Turbulence". He regarded the late 1930s as the foundational period when "the analytical and experimental means for the study of isotropic turbulence were clearly defined" but no effort was made to derive "the laws for the shapes of either the correlation or the spectral functions" (von Kármán 1948, p. 364):

> I believe this is the principal aim of the period in which we find ourselves at present. Promising beginnings were made by Kolmogoroff (1941), Onsager (1945), Weizsäcker (1946), and Heisenberg (1946). I do not want to follow the special arguments of these authors. Rather, I want to define the problem clearly and point out the relations between assumptions and results.

Kármán derived these results from a more general formalism. The $k^{-5/3}$-law for the spectral distribution of turbulent energy as a function of the wave-number k, for example, was merely a consequence of dimensional considerations. "That is the reason why it was independently found by Onsager (1945), Kolmogoroff (1941) and Weizsäcker (1946) (von Kármán 1948, p. 367)". Batchelor's and Kármán's papers show that the "remarkable series of coincidences" became within few years subject of authoritative reviews and scrutiny.

The agreement between these theories also fostered an optimistic attitude with regard to the fundamental questions of the nature of turbulence. Heisenberg expressed

with his "Remarks on the turbulence problem", published in June 1948, the opinion
that the basic physics of turbulence was no longer a mystery[3]:

> In earlier years Sommerfeld often discussed with his pupils the turbulence problem which was
> regarded for a long time as the unsolved fundamental problem of the newer hydrodynamics.
> In the meantime the statistical theory of turbulence founded about ten years ago by Taylor
> and v. Kármán has made so decisive progress that the turbulence problem in its physical core
> may arguably be regarded as solved.

The statistical theory of turbulence also transcended the border to neighbouring
disciplines. In 1949 IUTAM co-organized a conference with the International Astronomical Union (IAU) on "Problems of cosmical aerodynamics," held in August
1949 in Paris. The "present-day developments in hydro- and aerodynamics" were of
considerable importance for astronomers, the organizers announced in the preface,
particularly those on shock waves and turbulence. The latter was the subject of extensive presentations by Kármán and Weizsäcker (IUTAM and IAU 1949, pp. 129–148,
158–162, 200–203). John von Neumann used the Paris conference as an occasion
to meet with European colleagues in order to gain a first-hand impression on the
recent advances in the theory of turbulence. His European travels were funded by the
Office of Naval Research. "These trips were undertaken in connection with Contract
N7-onr-388, and, more specifically, with the work performed under that contract in
various fields of hydrodynamics and of the application of high-speed computational
methods to that subject and related ones," he remarked about the underlying interest
for this inquiry. "An important aspect of this program embraces various forms of
the theory of turbulence." Neumann did not regard his report to the Office of Naval
Research, written in 1949, as suitable for publication. Nevertheless the editors of
his collected works included it in the final volume, published in 1963, because it
was by that time "considered by a number of workers in the field to be one of the
most illuminating discussions of turbulence extant" (von Neumann 1949, p. 437).
(The rise of computational turbulence modelling will be subject of closer scrutiny
(see Sect. 6.1). Here Neumann's report is mentioned only to illustrate the surge of
interest in turbulence in the wake of the recent advances of the statistical theories.)

The excitement about the recent theories of turbulence prevailed for several more
years. In 1950 Dryden presented an invited address to a conference on fluid mechanics about "The Turbulence Problem Today". He came to the conclusion "that the
understanding of the turbulence problem has made some marked advances in recent
years" (Dryden 1951, p. 12). What he perceived as turbulence problem was no longer
confined to a specific challenge like in the early 1920s when Noether and others
regarded the frustrated efforts to explain the onset of turbulence in terms of hydrodynamic stability theory as the problem. Nor did he discern "the great problem of
developed turbulence" as the major big riddle, like Prandtl in his first papers on
the mixing length approach (see Sect. 2.3). Dryden's turbulence problem referred
to a whole bunch of research areas in which he noted advances during the past few
years. The pertinent references extended over several pages and were grouped under

---

[3] Heisenberg (1948a, p. 434). Translation ME.

the headings "Instability of Laminar Flow and Transition", "Isotropic and Homogeneous Turbulence", "Spectrum of Turbulence", "Diffusion", "Hot-Wire Anemometry", "Turbulent Boundary Layer", "Turbulent Wakes and Jets", "Turbulent Flow in Pipes and Channels", "Mixing Length" and "General Theory and Miscellaneous" (Dryden 1951, pp. 13–20).

Kármán's disciple Hans Liepmann published in November 1952 another survey on "Aspects of the Turbulence Problem". He introduced the review with the observation that turbulence affects a host of quite different fields. "The vast majority of the motions of fluids are turbulent. Consequently turbulence appears in many physical and astrophysical situations, sometimes as the prime mover and sometimes as a source of perturbation and of 'noise' in a general sense." In addition to the frequently exposed problem areas he addressed issues which had just recently become subject of investigation, like the impact of turbulence on wave propagation. Turbulent fluctuations of the refractive index, for example, suggested optical methods for turbulence investigations. Liepmann referred in this context to a recent study of Leslie Kovasznay who used shadowgraphs in order to determine the correlation of turbulent density fluctuations of high-speed air flows (Liepmann 1952, pp. 321, 334–336). Kovasznay had joined in 1948 the newly established Aeronautics Department of Johns Hopkins University in Baltimore, directed by another disciple of Kármán, Francis Clauser. Like Liepmann, Kovasznay represented a new generation of American scientists who regarded turbulence research as their major research field. Kármán counted Kovasznay for this research field "among the three best I know in this country, and among five or six I know of in the whole world."[4]

5.3 Turbulence as a Challenge for American Physics

The turbulence problem, as reviewed by Dryden and Liepmann, pertained to a host of practical applications. The investigations in Dryden's and Kármán's institutes had already been part and parcel of the scientific wartime effort in World War II. This legacy resulted in a reappraisal of fluid dynamics in the Cold War—especially in the USA. Against the background of their wartime experience, physicists in the USA argued that fluid dynamics in general, and turbulence in particular, deserved to be considered not only from the vantage point of engineering but also of physics. "Many of us feel that Fluid Dynamics is a subject of considerable physical interest which has been neglected by physicists in the past and revived only during the war." This is how an expert on shock waves complained after the war. "We are eager, therefore, to continue this interest, perhaps by having a permanent committee or even a division of the American Physical Society."[5]

---

[4]Kármán to Robert H. Roy, 29 December 1949. TKC, 16–29.
[5]Raymond Seeger to Hans Bethe, 28 August 1946. DFDA, 4-2.

A year later the American Physical Society (APS) announced the foundation of a new Division of Fluid Dynamics (DFD). Its major activity was to organize meetings that should be convened together with the annual APS conferences. Already the first DFD-meetings reflect the growing interest in turbulence. At the APS conference, held in January 1949 in New York, turbulence was made subject of a two-day symposium. The lecturers were, among others, Stanley Corrsin ("Some measurements in a round turbulent jet"), Kármán and Lin ("Statistical theory of isotropic turbulence"), Kovasznay ("Optical methods of measuring turbulence") and Martin Schwarzschild ("Turbulence in the atmosphere of stars").[6]

The DFD regarded the event so successful that turbulence entered the agenda of future meetings as a subject of special sessions. In summer 1949, for example, at the inauguration of new facilities at the Naval Ordnance Laboratory in White Oak, Maryland, the DFD convened another turbulence meeting—with Burgers as chairman. The names of the invited speakers and their themes show that turbulence was addressed in a much broader and more general manner than the site of this conference would have suggested: Kampé de Fériet lectured on the "Spectral tensor of a homogeneous turbulence"; Batchelor on "The nature of turbulent motion at large wave-numbers"; the astrophysicist Subrahmanyan Chandrasekhar on the "Development of Heisenberg's theory of the decay of isotropic turbulence"; Herbert Brian Squire on the "Investigation of the turbulence characteristics of an experimental low-turbulence wind tunnel"; Frenkiel made "Some remarks on turbulent diffusion".[7]

Batchelor used this occasion to promote Kolmogorov's theory in the United States.[8] His role in this regard can hardly be exaggerated. "When you are in Cambridge, you should try to see Batchelor, he is an awfully good man and very sound contrary to the other Cambridge luminaries in astrophysics," Chandrasekhar advised Martin Schwarzschild in May 1950.[9]

The rise of interest in turbulence in the USA also showed up during a DFD-meeting in November 1954 at Fort Monroe, Virginia.[10] Hugh Dryden delivered a banquet speech on "Fifty years of boundary-layer theory and experiment" which he concluded with some statistics on this topic. Although Dryden's focus was on boundary layers, the turbulence problem was a recurrent theme. "The origin of turbulence was the key theoretical problem," Dryden remarked for example about the trend of boundary layer investigations from "the second decade" of this fifty-years-history. In the subsequent decades the turbulence problem connected to boundary layers appeared in many different guises, such as "the heat-transfer problem" which in Dryden's statistics

---

[6]Proceedings of the American Physical Society, Minutes of the 1948 Annual Meeting at New York, January 26–29, 1949, Phys. Rev. 75 (1949), pp. 1279–1336.

[7]DFDA, 1–3 and 4–7.

[8]Smelt to Clauser, 13 April 1949, Ferdinand Hamburger Archives, Sheridan Libraries, Johns Hopkins University. Record Group 06.080, Department of Aeronautics, Subgroup 2, Series 1 (Professional Papers of Francis H. Clauser, 1946–1961, Folder: Smelt, Ronald, 1949).

[9]Chandrasekhar to Schwarzschild, 1 May 1950, Special Collections Research Center, University of Chicago Library, Subrahmanyan Chandrasekhar Papers 1928–1995, Box 28, Folder 12.

[10]DFDA, 1–3.

was represented by about 70 papers. In the mid-1950s, the number of boundary layer investigations was further growing. "The current total rate of production of papers is about 10 papers per month, nearly 9 times the rate immediately preceding World War II," Dryden concluded his survey (Dryden 1955).

By the mid-1950s the DFD also initiated the foundation of a new journal in order to provide space for the growing number of research papers on fluid dynamics. Previously such papers had been submitted either to engineering journals or to the *Journal for Applied Physics* which the DFD had originally considered as their preferred outlet. "There is at present no journal in the United States in which fundamental contributions to the field of fluid physics can be published with a view of circulation among those interested in the physics rather than the engineering aspects," DFD-representatives demanded in 1956 the foundation of a "Journal of Fluid Physics."[11] The authors of this memorandum were turbulence expert Frenkiel, who served as Secretary of the DFD, and geophysicist Walter M. Elsasser. By the same time Batchelor initiated in Great Britain the foundation of the *Journal of Fluid Mechanics*. He was not amused about the rival plans in the USA. "Do you not think that the Fluid Dynamics Division would be doing science a service by strengthening an international Journal rather than by setting up another national journal?", Batchelor attempted to prevent the competition.[12] But among the representatives of the American Physical Society there was "unanimous agreement" that both new journals would flourish.[13] Frenkiel argued that the competition between both journals would be "of a constructive nature."[14] The scope of *The Physics of Fluids*, as the new journal was titled, would include[15]:

> hydrodynamics, dynamics of compressible fluids, shock and detonation wave phenomena, hypersonic physics, rarefied gases and upper atmosphere phenomena, transport phenomena, hydromagnetics, ionized fluid and plasma physics, liquid state physics, superfluidity, boundary layer and turbulence phenomena, as well as certain basic aspects of physics of fluids bordering geophysics, astrophysics, biophysics and other fields of science.

As expected, both journals flourished. As their editors, Batchelor and Frenkiel, had contributed to turbulence research themselves with pioneering work, turbulence ranked high on the list of topics. In the four years from 1958 to 1961, both journals published 76 papers concerned with turbulence (35 in *The Physics of Fluids*, 41 in the *Journal of Fluid Mechanics*). As Frenkiel had predicted, there was no dearth of papers.

[11]Proposal for a Journal of Fluid Physics, 10 May 1956, AIP, Center for History, Physics of Fluids Records, Folder: Correspondence prior to 2 January 1958.
[12]Batchelor to Frenkiel, 7 February 1957, ibid.
[13]Minutes of the Meeting of the AIP Planning Committee, 21 February 1957, ibid.
[14]Frenkiel to Batchelor, 13 May 1957, ibid.
[15]Draft, 14 May 1957, ibid.

5.4   The First Textbooks on Turbulence

The surge of research papers on turbulence after World War II nourished the demand
of reviews and textbooks. Reviews like the FIAT reports on German wartime
research (Görtler 1948; Prandtl 1953) were surveys of diverse applications rather
than textbook-like presentations. In France the recent advances in the theory of turbulence became the subject of a series of lectures at the Sorbonne in February-March
1949. They were circulated under the title "Les Théories de la Turbulence" in a special series by the French Air Ministry. Subsequently, Alexandre Favre added some
recent experimental material. In this form the text was translated into English and
published as a NACA Technical Memorandum (Agostini and Bass 1950). In the same
line, the NACA had published in the preceding year Schlichting's wartime lectures
on boundary layer theory (Schlichting 1949a, b).

   Although these were textbook-like presentations on turbulence research, they
hardly fulfilled the need for an authoritative account. At a GAMM-meeting in 1951
Tollmien found it quite amazing that the turbulence problem has not become subject
of a systematic review despite remarkable progress in many areas. He criticised
the reports on turbulence at the preceding IUTAM-congresses in Paris and London
because they provided only glimpses at one or another aspect and did not attempt to
survey the problem of turbulence as a whole (Tollmien 1953).

   Tollmien was not the only expert who missed a systematic evaluation of the
turbulence problem in view of the recent advances. Batchelor regarded the "surge of
progress" since Kolmogorov's 1941 papers so remarkable and the need of students
for a coherent presentation so urgent that he found "the time is suitable for the
appearance of a book on homogeneous turbulence" (Batchelor 1953). It was the
right moment for such a monograph when the mood of progress was still in full
swing. As one of Batchelor's early students recalled, "*The Theory of Homogeneous
Turbulence*, published in 1953, appeared at a time when he was still optimistic that
a complete solution to the problem of turbulence might be found" (Moffatt 2002, p.
19). In view of Batchelor's role as a missionary of Kolmogorov's theory he was also
sufficiently motivated for this task. Chandrasekhar praised Batchelor's papers on the
theory of turbulence in 1951 as "the most satisfactory way to go about learning the
subject."[16]

   Several years later, Chandrasekhar's admiration of Batchelor waned. He criticised
"the sterility of his approach to scientific problems".[17] There may also have been
some rivalry because he planned himself a textbook on "The Statistical Theory of
Turbulence", as he had confessed to a colleague in August 1950. The plan matured
into a book proposal to Clarendon Press, but when Batchelor's monograph appeared
in 1953, the need for another monograph on the same subject vanished. "I have since
changed my mind and would rather write my book on stability first," Chandrasekhar

---

[16]Chandrasekhar to Taylor, 24 September 1951, Special Collections Research Center, University
of Chicago Library, Subrahmanyan Chandrasekhar Papers 1928–1995, Box 31, Folder 3. See also
Moffatt (2011).

[17]Chandrasekhar to O. J. Eggen, 13 September 1968, ibid., Box 11, Folder 11.

informed the publisher.[18] Chandrasekhar's *Hydrodynamic and Hydromagnetic Stability* appeared at Clarendon Press in 1961 and became a classic. His originally planned textbook on the statistical theory of turbulence never appeared.

Despite Chandrasekhar's reproach of sterility, Batchelor's *Theory of Homogeneous Turbulence* may be rated as the first authoritative textbook on uniform turbulent flows, such as wind tunnel turbulence behind grids. Three years later Alan Townsend, Batchelor's close colleague at Cambridge, published *The Structure of Turbulent Shear Flow* (Townsend 1956). (On Townsend as a pioneer of postwar turbulence research see Marusic and Nickels (2011).) It may be considered the first textbook on non-uniform turbulence, such as boundary layer turbulence or the turbulent flow in jets, pipes and channels. In view of its wide scope a reviewer suggested that "it might have been called simply *Turbulence*" (Lighthill 1956, p. 555).

The first textbook with this general title was published another three years later by J. O. Hinze, a professor at the Technical University in Delft. It originated from lectures for engineers of the Shell laboratories in Amsterdam and Delft. In view of this audience "special attention has been paid to discussing the mechanism of turbulence in relation to flow resistance and to heat and mass transfer... a complete textbook on turbulence was not intended" (Hinze 1959, p. v). The content of *Turbulence* was limited to fully developed turbulence and dismissed the problem of the transition from laminar to turbulent flow. Yet the treatment started from first principles and did not shy away from the basic concepts of the statistical theory of turbulence as founded by Taylor, Kármán, Kolmogorov and others.

5.5   Marseille 1961

In 1960 representatives of the International Union of Geodesy and Geophysics (IUGG) and IUTAM conceived the plan of a common symposium on turbulence. Like the joint ventures of IUTAM with the International Astronomical Union the symposium with the IUGG was supposed to focus on the role of turbulence in neighbouring sciences, in this case meteorology, oceanography and other geophysical disciplines. The plan met with Alexandre Favre's initiative to establish an Institute for the Statistical Mechanics of Turbulence at Marseille. As a result, Marseille became in summer 1961 the site of an extraordinary double-event in the history of turbulence. From 28 August to 2 September Favre celebrated the foundation of his institute with an international colloquium on "The Mechanics of Turbulence," and in the subsequent week, from 4 to 9 September, the IUGG and IUTAM held a common international symposium on "Fundamental Problems in Turbulence and Their Relation to Geophysics."

Favre opened the colloquium at his new institute with the remark that "the Mechanics of Turbulence can now be considered as a new science". The goal of the colloquium was "a critical examination of works published on the study of turbulence"

---

[18]Chandrasekhar to Clarendon Press, 10 December 1956, ibid., Box 68, Folder 8.

(CNRS 1962, p. 5). The proceedings amounted to a volume of 470 pages. The contributions to the subsequent IUTAM-IUGG symposium were printed separately in the *Journal of Geophysical Research*. Altogether the list of contributors at both events reads like a Who is Who of contemporary research on turbulence, with a strong participation of young scientists from the USA who would soon dominate the international research front. At the height of the Cold War it is furthermore remarkable that the Iron Curtain seemed to be permeable temporarily for an elite from the Moscow Academy of Science: Western researchers had an opportunity to meet for the first time Kolmogorov and a few of his fellow academicians whose names were known outside Russia only from Batchelor's publications or a recent German translation of Kolmogorov's papers which had appeared in the German Democratic Republic (Goering 1958).

For the organisers of these events it was obvious that their "new science" required a critical evaluation of what was to be regarded as the turbulence problem, or, in Favre's words, "the fundamental and general ideas concerning the mechanism and the structure of turbulence" (CNRS 1962, p. 6). The preparations for Favre's colloquium had started a year earlier by the definition and categorisation of the turbulence problem into seven branches on the following topics:

Diffusion and Lagrangian effects (Corrsin)

Energy transfer in homogeneous turbulence (Batchelor)

Steady fully developed turbulence (R. W. Stewart)

Free turbulence (Liepmann)

Turbulent boundary layers (Liepmann and Schlichting)

Turbulence in compressible and electrically conductive media (Kovasznay)

New concepts and recent contributions (Joseph Kampé de Feriét)

The chosen representatives of each branch acted as chairmen of the respective sessions. They also selected the topics and speakers of individual presentations. An eight's session was dedicated to the presentation of summaries from each branch.

Among the branches of turbulence research those on "Energy transfer in homogeneous turbulence" and on "Steady fully developed turbulence" were particularly remarkable. Batchelor had promoted Kolmogorov's theory (K41) so successfully that it was widely regarded as one of the few solid pillars for a theory of fully developed turbulence. This view received further support when Stewart, who chaired this session, presented slides of the energy spectrum of turbulence from velocity measurements in a tidal channel which corroborated K41 in a most convincing manner (Grant et al. 1962).

But the euphoric mood did not last long—because Kolmogorov cast doubt on the underlying assumptions of his old theory: K41 ignored spatial fluctuations in the rate of turbulent dissipation of energy. Kolmogorov's Marseille lecture opened the door to studies of intermittency, a kind of spottiness in the statistics of the turbulent energy transfer. "This lecture, Kolmogorov's final contribution to turbulence, was, to put it mildly, a bit of a bombshell!" (Moffatt 2012, p. 3).

It exceeds the scope of this brief historical account to review the achievements presented in the seven sections of the Marseille 1961 event. Most speakers probably agreed with Favre's opinion that they all witnessed the rise of turbulence as a new science. One participant aptly called "Marseille 1961: A Watershed for Turbulence" (Moffatt 2002, p. 21). Its impact was crucial and is even felt more than fifty years later. "Turbulence remains one of the oldest and most challenging research problems in both pure and applied science," the organisers of the "Turbulence Colloquium Marseille 2011" introduced the event that celebrated the fiftieth birthday of the Marseille Colloquium 1961. "This Colloquium led to the development of research areas that are still very much alive to this day" (Farge et al. 2011, p. 1).

Chapter 6
Computational Approaches

Abstract The advent of the electronic computer opened new approaches to the turbulence problem. As early as in 1946 John von Neumann discerned turbulence as a
challenge for the Electronic Computer Project at the Princeton Institute for Advanced
Study (IAS). In the 1950s the IAS-computer became a role model of a first generation of digital high speed computers. The Cold War fuelled not only the development
of computers but also the birth of computational sciences—first at Los Alamos and
other facilities where research on atomic bombs and other weapons entailed a host of
problems that could be solved by computational means only. Another area of rapid
growth was numerical weather forecast, where atmospheric turbulence became a
major problem. Concepts like Large Eddy Simulation (LES) were developed in order
to compute flows on geophysical scales. The small unresolved scales required "sub-
grid" modelling. Turbulence models based on the decomposition in mean and fluctuating quantities, however, had to address the "closure problem" because they resulted
in more unknowns than equations. Until computers would be powerful enough for
Direct Numerical Simulation (DNS), modelling of turbulent flows by one or another
closure method remained the only viable computational approach.

"I have always believed," John von Neumann wrote in 1946 to a colleague, "that very
high speed computing could replace some—but of course not all—functions of a wind
tunnel." Neumann referred in this letter to the wind tunnel as an "analogy" computer
in contrast to the "electronic digital computer" which he had just begun to develop
with a team of engineers and applied mathematicians at the Princeton Institute for
Advanced Study. It was supposed "to handle problems of very high complexity—
actually of much higher complexity than that of the typical wind tunnel—or flow-
problems..." Alluding to the giant wind tunnels at the large aeronautical facilities in
the USA he added "that such a machine would be much smaller and cheaper than a
conventional wind tunnel."[1]

Thus the use of numerical means for solving flow problems was debated even
before the first digital computers entered the scene. But even fifteen years later, the

---

[1] John von Neumann to Howard Emmons, 3 April 1946, quoted in Aspray (1990, p. 274).

© The Author(s), under exclusive license to Springer Nature Switzerland AG 2019   61
M. Eckert, *The Turbulence Problem*,
SpringerBriefs in History of Science and Technology,
https://doi.org/10.1007/978-3-030-31863-5_6

turbulence problem appeared inaccessible by such means. Stanley Corrsin, a rising star among the turbulence researchers in the USA, summarized in 1961 the research on "Turbulent Flow" in the magazine *American Scientist* in this way: "At present, we have a qualitative understanding of the phenomenon, and can even predict some of its features quantitatively from Newton's Laws. But much of the core of the turbulence problem has yet to yield to formal theoretical attack." By this time the electronic computer had already been developed into a new tool for solving problems in fluid dynamics numerically. "In closing, we must certainly speculate on the future role of large computing machines in turbulence research. Valuable computations have already been made," Corrsin concluded his review. But a rough estimate about the computational requirements to simulate three-dimensional turbulence at high Reynolds numbers down to the Kolmogorov microscale left him pessimistic: "The foregoing estimate is enough to suggest the use of analog instead of digital computation; in particular, how about an analog consisting of a tank of water?" (Corrsin 1961, pp. 300, 324).

6.1 John von Neumann and the Electronic Computer Project

As an advisor of the atomic bomb project at Los Alamos, New Mexico, and the Ballistic Research Laboratory (BRL) at Aberdeen, Maryland, Neumann was fully aware of the role of computational tools for the development of new weapons. But he envisioned more uses than just weapons. While the legendary "Electronic Numerical Integrator and Computer (ENIAC)" was designed for the computation of firing tables and still being developed, Neumann planned a high-speed computing machine for general purposes[2]:

> The use of a completed machine should at first be purely scientific experimenting [...] It is clear now that the machine will cause great advances in hydrodynamics, aerodynamics, quantum theory of atoms and molecules, and the theory of partial differential equations in general. In fact I think that we are still not able to visualize even approximately how great changes it will cause in these fields. [...] Finally it should be used in applied fields: it will open up entirely new possibilities in celestial mechanics, dynamic meteorology, various fields of statistics, and in certain parts of mathematical economics, to mention only the most obvious subjects.

In order to materialise this vision Neumann turned to Carl-Gustav Rossby, an expert on atmospheric flows, for advice on how to make use of the computer for meteorology. Rossby suggested to focus at first on the equations for the large-scale circulation on the globe, as a prerequisite for future numerical weather forecast. He proposed to form a small team of theoretical meteorologists. By summer 1946 it was clear that meteorology would be a major part of the Electronic Computer Project at the IAS (Harper 2008, Chap. 4). Rossby had already participated in the turbulence

---

[2]John von Neumann, Memorandum, 5 September 1945, quoted in Aspray (1990, p. 52).

symposium at the Fifth International Congress for Applied Mechanics in 1938 at Cambridge, Massachusetts, with lectures on turbulent mixing in the atmosphere. In his discussions with Neumann about the role of the computer for numerical weather forecast turbulence must have surfaced as a major problem. When Neumann delivered in May 1946 a speech before a Navy agency which would fund his Electronic Computer Project during the following years, turbulence was explicitly exposed as a major problem[3]:

> Clearly one of the major difficulties of fluid dynamics, which turns up at the most varied occasions, is the phenomenon of turbulence. The major reasons why we cannot do much about it analytically are that it involves a nonlinear, partial differential equation (and it is really nonlinear; you lose the decisive phenomena if you attempt to linearize it), and that it is quite intrinsically three-dimensional. [...] In addition, a fourth dimension (time) has to be added to these three, because turbulence is necessarily a transient, nonstationary phenomenon. [...] I don't say that the problems of turbulence are necessarily the most important ones to be solved by fast machine calculation, but I would rather expect that many other problems that one will want to solve with such machines will prove to have a lot in common with the general situation that exists in the case of turbulence, and turbulence happens to be a significant and relatively familiar example.

Thus turbulence played from the very beginning a major role as a paradigmatic challenge for Neumann's computer program. "An important aspect of this program embraces various forms of the theory of turbulence," he introduced a review on turbulence three years later, when the IAS-computer was still under development. The review resulted from a trip through Europe which Neumann used to meet the leading experts and to discuss all aspects of the turbulence problem. He was conscious that the computational approach would require "quite advanced facilities", but that did not prevent him to keep turbulence as an important item on the agenda of his "high-speed computing program". It deserved particular attention "in connection with the theoretical problems of turbulence" (von Neumann 1949, p. 438).

6.2 Early Numerical Solutions of the Stability Problem

"The problems which lead to turbulence" ranked on top of Neumann's agenda. He found it "very plausible that turbulence is a phenomenon of instability". Turbulent flow, from this perspective, "represents one or more solutions of a higher stability" acquired at higher Reynolds numbers. Instead of one "turbulent solution" of the Navier-Stokes equation, Neumann envisioned a multitude of solutions characterized by statistical properties in which "the essential and physically reproducible traits of turbulence" would be manifest. The study of turbulence, in Neumann's view, should proceed in two steps (von Neumann 1949, pp. 438–440):

---

[3]Neumann, quoted in Stern (1981, p. 267).

Firstly, the stability properties of the laminar flow must be investigated. Secondly, there
is need of a complete theory of the common statistical properties of large, statistically
homologous families of solutions, which exhibit the characteristic turbulent traits. Thus,
the theory immediately divides into two halves: That of the stability theories and that of the
statistical theories.

Therefore it is not accidental that Lin's recent review of the Orr-Sommerfeld
approach attracted Neumann's attention as a starting point for the first step. At the
suggestion of Lin, Neumann asked Chaim Pekeris, a former student of Rossby and
one of the first collaborators in Neumann's Electronic Computer Project, to make the
case of Heisenberg's doctoral thesis, i. e. the solution of the Orr-Sommerfeld equation
for plane Poiseuille flow, subject of a numerical study. Pekeris was the ideal candidate
for this task. In the 1930s, he had chosen the disputed theory of hydrodynamic
stability to put to the test his skills as an applied mathematician (Pekeris 1936,
1938). While the computer at the IAS was still under development, Neumann and
Pekeris resorted to the Selective Sequence Electronic Computer (SSEC) at the IBM
Watson Laboratory at Columbia University. The SSEC was the first machine which
could perform electronic computations by a stored program (Bashe 1982). It extended
along three sides of a hall "60 feet long and 20 feet wide so that visitors actually
stood inside the computer," as a history of the IBM Watson Laboratory remarked
(Brennan 1971, p. 21).

By 1950 Pekeris had prepared the problem for a first test run at the SSEC, but the
result "did not settle the question at issue," Llewellyn H. Thomas later commented
these beginnings. Thomas had been recruited as an applied mathematician at the
Watson Laboratory and pursued the problem when Pekeris left Neumann's group
in order to start electronic computer development in Israel. Two years later Thomas
reported success. In a preliminary communication he concluded that "it may now be
regarded as proved that plane Poiseuille flow becomes unstable at about $R = 5800$"
(Thomas 1952, p. 813). In his final paper he presented more details about these computations (Thomas 1953). They may be regarded as the first computational solution
of what was recognized as *the* turbulence problem three decades ago. By the same
token, the numerical results justified the asymptotic methods with which Lin had
improved Heisenberg's approach (see Sect. 4.3). The computational results settled
the dispute about Heisenberg's and Lin's results. Lin further consolidated the theory
with a review on hydrodynamic stability (Lin 1955) that served Heisenberg several
years later "as an indication that approximation methods derived from physical intu-
ition are frequently more reliable than rigorous mathematical methods, because in
the case of the latter it is easier for errors to creep into the fundamental assumptions"
(Heisenberg 1969, p. 47).

Neumann's IAS-computer (Fig. 6.1) became a role model for similar computers
in the USA and abroad. Electronic computers from this generation were expensive
facilities; it is therefore not astonishing that early numerical investigations of hydrodynamic stability originated at research centres like the Jet Propulsion Laboratory or
The Northrop Aircraft Company–Cold War facilities which could afford such computers (Brown 1959; Mack 1960). With the rise of Computation Centres elsewhere,

Fig. 6.1 John von Neumann (right) next to J. Robert Oppenheimer, director of the Princeton Institute for Advanced Study, in October 1952 at the dedication of the IAS computer

"Computer-Aided Analysis of Hydrodynamic Stability" (Kurtz and Crandall 1962) emerged as a third research mode besides theory and experiment.[4]

By the same time, new experiments (Klebanoff et al. 1962) and theoretical studies (Stuart 1962) suggested a scenario for the transition to turbulence which was far beyond the reach of the Orr-Sommerfeld approach. Direct numerical flow simulations at Los Alamos about the vortex formation in the wake of an obstacle reproduced the scenario known from laboratory experiments: a transition from laminar steady flow via a "Kármán vortex street" to a turbulent wake at high Reynolds numbers (Harlow and Fromm 1963). Although these numerical experiments were only two-dimensional, the similarity with the vortex formation in laboratory experiments raised the hope that the computer could also become a tool for solving more fundamental problems.

For the time being, however, computational investigations of hydrodynamic stability did not lay the riddle of the onset of turbulence to rest. The demarcation between stable and unstable states of flow could indicate the initiation of the transition to turbulence, but not the processes that occurred within the transition zone. No theory could account for this complexity. "Experiment and theory agreed, as far as eigenvalues and eigenfunctions were concerned," the authors of a textbook on *Stability of Parallel Flows* summarized the research front in their field by the mid 1960s. "Yet turbulent transition was not understood, and it still remains an enigma" (Betchov and William O. Criminale 1967, p. 3). In a 1968-review "On the many faces

---

[4]There is a rich literature on "computer experiments" and "computer simulation" as a new mode of research; for an overview from an epistemological perspective see Winsberg (2010).

of transition" the speaker conjectured "that many instability paths to turbulence are admissible". In view of this complexity the prediction of transition was "a peculiarly nondeterministic problem" (Morkovin 1969, p. 2).

The rapid development of electronic digital computers made the problem of laminar-turbulent transition a recurrent item on the agenda of applied mathematicians, physicists and engineers with access to computing centres. Yet the "bewildering variety of transitional behavior" (Morkovin 1969, p. 1) eluded the computational capabilities at least for another decade. In 1979 a first symposium on "Laminar-turbulent transition" was held in Stuttgart under the umbrella of IUTAM. Subsequently IUTAM sponsored other symposia under the same title in Novosibirsk (1984) and Toulouse (1989), to name only those which consolidated this research field as an ongoing concern for IUTAM.[5] The symposium in Toulouse was opened with a "Dialogue on Progress and Issues in Stability and Transition Research" presented by two experts who had contributed to this research field for more than thirty years. With the focus on the more recent developments since the first IUTAM Symposium in 1979 they expressed their belief "that there is no universality of the evolutionary path to turbulent flow even in geometrically similar mean laminar shear flows." They acknowledged "major developments in the tools of transition research: experiment, analysis and computation" and regarded the latter "now a full partner in our threepronged attack on the understanding of transition." But even contemporary supercomputers like CRAY II added little to solve the fundamental riddles of transition. The "formulation of computational approaches" remained an issue for the future (Morkovin and Reshotko 1989, pp. 3–4).

6.3   The Origins of Large-Eddy Simulation

Although stability theories ranked first in John von Neumann's survey on "Recent Theories of Turbulence", he certainly considered fully developed turbulence as a problem that deserved more attention—not the least because "it plays a decisive role in the momentum as well as the energy exchanges of the terrestrial atmosphere and the oceans—that is, in meteorology and in oceanography," as he remarked in his survey under the headline "Computational Possibilities" (von Neumann 1949, p. 467). The intended use of the IAS-computer for numerical weather forecast involved the consideration of atmospheric turbulence. Neumann and his Meteorology Group at the IAS not only pioneered the simulation of large-scale atmospheric circulation but also gave an impact on what became known as Large Eddy Simulation (LES) of turbulence.

Computational fluid dynamics on coarse grained grids had to cope in general with the problem of the smaller scales below the resolution of the finite elements of the mesh—not only in turbulence. Besides the problem of maintaining numerical stability it was not clear how to deal with the unresolved sub-grid scales. The computation

---

[5]http://www.iutamtransition2019.org/history/ (26 October 2019).

of shock waves, for example, was enabled by the smoothing effect of an "artificial viscosity" which was added to the basic equations (von Neumann and Richtmyer 1950). In turbulence, the eddy viscosity appeared as the appropriate candidate for relating sub-grid scales to the numerically resolved grid. "The lateral transfer of momentum and heat by the non-linear diffusion, which parametrically is supposed to simulate the action of motions of sub-grid scale, accounts for a significant portion of the total eddy transfer." This was the way how turbulence was included by Joseph Smagorinsky in an early model about atmospheric circulation (Smagorinsky 1963, p. 99). It marks the beginning of LES.

Smagorinsky was a meteorologist at the U.S. Weather Bureau in Washington, D.C., and his model represents the first effort to derive the main features of the general atmospheric circulation directly from the Navier-Stokes equations. "The present study is an attempt to employ the primitive equations for general circulation experiments," Smagorinsky declared about the goal of his work. It was "an outgrowth of collaboration with J. G. Charney, N. A. Phillips, and J. von Neumann, who engaged in the initial planning stages of this investigation" (Smagorinsky 1963, p. 100). Thus Smagorinsky acknowledged his affiliation with Neumann's group at Princeton. "In 1949, I was invited as an occasional visitor, from my base in Washington, D.C., to assist the group in extending its one-dimensional linear barotropic calculations", he recalled later his first encounter with Neumann's group. "On behalf of the Weather Bureau, I also was asked to become familiar with the theoretical aspects of a more realistic model" (Smagorinsky 1983, p. 7). Upon Neumann's suggestion the Weather Bureau established in 1955 under Smagorinsky's direction a General Circulation Research Section, the precursor of the Geophysical Fluid Dynamics Laboratory (GFDL) (Edwards 2010, Chap. 7). In contrast to numerical weather forecast, the GFDL aimed at models for the general atmospheric circulation pattern as developed in the long run, what Neumann called the "infinite forecast".[6]

It is therefore not accidental that Smagorinsky's effort to model turbulence was published in the *Monthly Weather Review*. The year before Smagorinsky's collaborator Douglas K. Lilly had published in another meteorological journal, *Tellus*, a study "On the numerical simulation of buoyant convection" which may be regarded as a precursor to Smagorinsky's model. "Doug did invent the essence of LES along the way!", Smagorinsky praised Lilly's early contributions to turbulence many years later (Kanak 2004, p. 3). In 1964 Lilly became employed as senior scientist at the National Center for Atmospheric Research (NCAR) in Boulder, Colorado. In November 1966 he presented a paper on "The Representation of Small-Scale Turbulence in Numerical Simulation Experiments" at an IBM Scientific Computing Symposium. Earlier approaches "lacked suitable mechanisms for simulating the development and maintenance of a three dimensional turbulent energy cascade", Lilly qualified the previous two-dimensional models. "The future practicality of such computations seems to

---

[6]Neumann in a "Proposal for a Project on the Dynamics of the General Circulation", reproduced in Smagorinsky (1983, p. 30): "Indeed, determining the ordinary circulation pattern may be viewed as a forecast over an infinite period of time, since it predicts what atmospheric conditions will generally prevail when they have become, due to the lapse of very long time intervals, causally and statistically independent of whatever initial conditions may have existed.".

require development of equations describing the transport of turbulent energy into and through the inertial range" (Lilly 1967, Abstract).

Three years later, James Deardorff provided solid evidence for the practicability of LES with "A numerical study of three-dimensional turbulent channel flow at large Reynolds numbers" (Deardorff 1970). Deardorff was Lilly's colleague at the NCAR. "Doug showed me how to finite-difference the vorticity and thermal-diffusion equations so as to conserve kinetic energy and temperature variance in the absence of sources and sinks," Deardorff recalled how Lilly introduced him to computational fluid dynamics at the NCAR[7]:

> After computing power had increased to the point where it was conceivable to study tur-
> bulence in three dimensions, in the late 1960s, the problem arose of how to simulate the
> dissipation of turbulent kinetic energy cascaded to scales too small to represent explicitly.
> Doug was very well acquainted with J. Smagorinsky's work on this subject, and was very
> helpful in advising me on how to apply Smagorinsky's method to small-scale turbulence.
> Doug had already done his own research on this problem, and so could recommend a coef-
> ficient of proportionality between the magnitude of the subgrid-scale eddy coefficient and
> the resolvable strain rate.

Although the focus at the NCAR was on atmospheric circulation, Deardorff was eager to demonstrate the practicability of LES more generally. Instead of publishing his study like Smagorinsky and Lilly in a meteorological journal, he chose the *Journal of Fluid Mechanics* and explicitly declared as his goal "to test this meteorological approach upon an interesting case of laboratory turbulence: plane Poiseuille flow (channel flow) driven by a uniform pressure gradient" (Deardorff 1970, p. 454). Atmospheric flow remained Deardorff's predominant interest, but turbulence played an ever growing role. Three years later he published a study on "The Use of Subgrid Transport Equations in a Three-Dimensional Model of Atmospheric Turbulence" in the *Journal of Fluids Engineering* (Deardorff 1973). "Numerical modeling of the details of fluid flow at large Reynolds numbers has progressed at a rate commensurate to the development of the digital computer," Deardorff introduced his model. It simulated the flow over a heated ground area of eight square kilometers and a height of 2 km, using a computational grid of $40 \times 40 \times 40$ points. The computation was executed on a supercomputer of the 1970s, the CDC 7600, designed by Seymour Cray. Another researcher with access to such supercomputers, Anthony Leonard from the NASA Ames Research Center, Moffett Field, California, used by the same time the label "large eddy simulation" for this approach (Leonard 1974).

Deardorff's approach became applied at other establishments which could afford the required computational facilities. Ulrich Schumann, a doctoral student at the Technical University Karlsruhe and collaborator at the Karlsruhe Nuclear Reactor Center, submitted in 1973 his dissertation on numerical investigations of turbulent flows in plane channels and concentric annuli. Schumann also averaged the Navier-Stokes equations over grid volumes, but modified Deardorff's model at the sub-grid level. His computations were performed on the mainframe computer of the Karlsruhe

---

[7]Quoted in Kanak (2004, pp. 5–6).

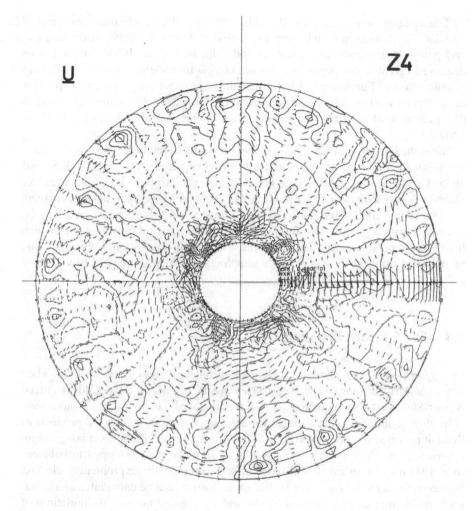

Fig. 6.2 Schumann's computation of turbulent velocities at one moment in an annular space. Arrows represent the velocity components in a plane perpendicular to the axis, contour lines the axial velocities (Schumann 1973, Fig. 17)

Nuclear Reactor Center, an IBM 370/165; typical computational times took several hours (Schumann 1973, 1975) (Fig. 6.2).

The LES-approach also entered other engineering establishments. In 1977, it was presented to the community of aeronautical science in the *AIAA Journal* (Ferziger 1977, Abstract):

Large eddy simulations are a numerical technique in which large-scale turbulent structures are computed explicitly, and the small structures are modeled. Arguments for believing this method to be superior to more conventional approaches are given, the basis of the method is given, and some typical results displayed. The results show that the method does have enormous promise, but much further development is required.

The computations were carried out like those of Deardorff four years earlier on CDC 7600 computers, with running times of 90 min for cubic grids with 64^3 grid points. "Large eddy simulation methods," the author concluded, "already have displayed a great deal of potential as important tools for understanding and predicting turbulent flows." Furthermore, LES raised "the hope of partially replacing expensive experimental work with less costly computation. Over the longer range, it is possible that LES methods will become a standard computational tool" (Ferziger 1977, p. 1267).

Since the late 1970s, LES was on the way towards "a really big industry".[8] Since the 1990s LES may be studied from textbooks (Galperin and Orszag 1993; Sagaut 2005; Grinstein et al. 2007). Even broader oriented textbooks on turbulence, like Stephen Pope's *Turbulent Flows* published in 2000, include comprehensive chapters on LES (Pope 2000, Chap. 13). The "big industry" in this research area may be estimated from a glimpse into the ISI science citation index from that year which listed 164 papers including the keywords "large-eddy-simulation". "By 2004 this number had doubled to over 320 per year Sagaut" (2005, Foreword to the Third Edition).

6.4   The Closure Problem

Strictly speaking, the closure problem of turbulence dates back to 1894 when Reynolds derived equations for fully developed turbulence: the Reynolds-averaged Navier-Stokes (RANS) equations. These are unclosed, i.e. they contain more variables than equations. The additional variables appear as time-averaged products of fluctuating velocities and are known as Reynolds stresses. Prandtl's mixing length approach from 1925 and his "new system of formulae for developed turbulence" from 1945 may be regarded as different efforts to cope with the problem of closure. However, these are retrospective evaluations. Closure became enunciated as a turbulence problem in its own right only in the wake of the first numerical simulations of turbulence. Invoking the appropriate "closure" raised among early turbulence modellers the hope "that the barricades against successful turbulence theories are finally beginning to crumble under the combined attack of empiricism, analytic rigor, and numerical simulation" (Fox and Lilly 1972, p. 52).

Once more it is not accidental that the protagonists in this area came from institutions that disposed of powerful computational means and were oriented to applied research like the National Center for Atmospheric Research (NCAR) in Boulder, Colorado. Large-Eddy Simulations of geophysical circulation on meshes where the distance between the grid points corresponded to hundreds of kilometers involved

---

[8] Anthony Leonard, Interview by Heidi Aspaturian. Pasadena, California, November 6, 9, 14, and 21, 2012. Oral History, http://resolver.caltech.edu/CaltechOH:OH_Leonard_A (26 October 2019). See the review articles (Rogallo and Moin 1984; Schumann and Friedrich 1985; Lesieur and Métais 1996; Meneveau and Katz 2000).

sub-grid modelling of turbulence which required closure equations that accounted for scales down to the Kolmogorov microscale of some millimeters. Turbulence modelling for industrial applications or in aeronautical research had to cope with scales from fractions of millimeters to several meters. In 1968 a conference was convened in Stanford on the computation of turbulent boundary layers with the goal to compare different closure methods. The applied context is illustrated by the sponsors of this event: the Mechanics Division of the U.S. Air Force Office of Scientific Research and the industrial sponsors of the Internal Flow Program of the Mechanical Engineering Department at Stanford University (Kline et al. 1969).

In the aftermath of this conference, William Craig Reynolds from the Department of Mechanical Engineering at Stanford University reviewed the "Recent Advances in the Computation of Turbulent Flows" in a course on turbulence for the American Institute of Chemical Engineers. By that time, in 1970, Reynolds already distinguished between a variety of closure models for different purposes. MVF (Mean Velocity Field) closure was regarded as appropriate for boundary layer flow. Another type was designated as MTE closure (Mean Turbulent Energy); it was found suitable in situations when the mean flow changed abruptly. Closure methods based on the RANS equations were labelled as MRS closure (Mean Reynolds Stress), but "such closures are not yet tools for practical analysis." Another type of closure was designated with the acronym FVF (Fluctuating Velocity Field). The review closed with an outlook to the future (Reynolds 1974, p. 242):

> It should not be long before simple boundary-layer flows are routinely handled in industry by MVF prediction methods. These methods are easy to use, require a minimum of input data, and give results which are usually adequate for engineering purposes. MTE methods will become increasingly important to both engineers and scientists, for they afford the possibility of including at least some important effects missed by MVF methods. The debate over the gradient-diffusion vs. large-eddy-transport closures will continue, and both methods will probably continue to be used with nearly equal success. MRS methods will be explored from the scientific side, but probably will not be used to any substantial degree in engineering work for some time to come.

It was obvious that closure methods were crucial for computational turbulence modelling in a host of applications. The rapid sequence of review articles illustrates the pace of progress in this field. In 1972, two NCAR-researchers summarized the state of turbulence modelling for geophysical flows, also "as a test for turbulence theory" (Fox and Lilly 1972). By the same time Peter Bradshaw, a professor of Experimental Aerodynamics at the Department of Aeronautics at the Imperial College in London, elaborated on "the closure problem" in a lecture on "The understanding and prediction of turbulent flow" with the focus on turbulence models suited for engineering purposes (Bradshaw 1972). In the same year the *AIAA Journal* presented to the community of aeronautical engineers "A Survey of the Mean Turbulent Field Closure Models" (Mellor and Herring 1972). In a textbook on turbulent boundary layer flow published in 1974 "as an outgrowth of work done primarily in the Aerodynamics Research Group at the Douglas Aircraft Company" a chapter reviewed the closure methods which had been developed so far for boundary layer turbulence (Cebeci and Smith 1974).

In 1976 William Craig Reynolds, by now chairman of the Department of Mechanical Engineering at Stanford University, updated his survey from 1970 for the *Annual Review of Fluid Mechanics*. "By the mid-1960s there were several workers actively developing turbulent-flow computation schemes based on the governing partial differential equations (pde's)," he remarked about the recent beginnings of turbulence modelling. "The first such methods used only the equations for the mean motions, but second-generation methods began to incorporate turbulence pde's." He distinguished the following turbulence models (Reynolds 1976, pp. 183–184):

1. Zero-equation models–models using only the pde for the mean velocity field, and no turbulence pde's.

2. One-equation models–models involving an additional pde relating to the turbulence velocity scale.

3. Two-equation models–models incorporating an additional pde related to a turbulence length scale.

4. Stress-equation models–models involving pde's for all components of the turbulent stress tensor.

5. Large-eddy simulations–computations of the three-dimensional time-dependent large-eddy structure and a low-level model for the small-scale turbulence.

A successful two-equation model from the second generation, for example, was the k-ε-model developed in the early 1970s at the Department of Mechanical Engineering of the Imperial College in London. Closure was achieved by two partial differential equations for the mean flow behaviour in which the dependent variables were the turbulence kinetic energy k and its dissipation rate ε. "It is the simplest kind of model that permits prediction of both near-wall and free-shear-flow phenomena without adjustments to constants or functions," the authors praised its virtues, "it successfully accounts for many low Reynolds-number features of turbulence; and its use has led to accurate predictions of flows with recirculation as well as those of the boundary-layer kind" (Launder and Spalding 1974, p. 287).

Turbulence modelling involved the consideration of specific flow configurations, empirically determined constants and computational economy rather than universal aspects of turbulence. Despite an explosive increase in computer power and algorithmic sophistication in Computational Fluid Dynamics (CFD) this has not changed over the subsequent decades. The closure problem remained crucial for turbulence modelling in a host of engineering applications. "Closure models continue to play a major role in applied CFD and remain an area of active research and development" (Durbin 2017, p. 77).

6.5 Direct Numerical Simulation (DNS)

The only way to avoid the closure problem was to solve the Navier-Stokes equations directly down to the smallest scales, i.e. without decomposing velocities and pressures in mean and fluctuating values. From the perspective of the 1960s such an approach was elusive. As Corrsin estimated in 1961, the computational grid would require at least 10^{12} points in order to resolve for a flow at a Reynolds number of 10^4 the turbulent eddies down to the Kolmogorov microscale. "The number of 'bits' of information is 2.7×10^{13}" (Corrsin 1961, p. 324), far beyond the computational capabilities at the time and in the foreseeable future. A decade later, an estimate for the direct numerical simulation of turbulent pipe flow arrived at about 10^9 s or 100 years for even a "somewhat scanty turbulence computation" (Emmons 1970, p. 33).

DNS remained prohibitive for many more years as a computational approach in the research of turbulent flows, not to mention engineering applications. With respect to computational economy, "Progress in the development of a Reynolds stress turbulence closure" seemed most rewarding (Launder et al. 1975)—to quote the title of "This Week's Citation Classic" from ten years later.[9] Computationally more expensive was LES with appropriate sub-grid modelling. "At the present time, simulation can provide detailed information only about the large scales of flows in simple geometries," a review remarked about the LES-approach in 1984, "and is advantageous when many flow quantities at a single instant are needed (especially quantities involving pressure) or where the experimental conditions are hard to control or are expensive or hazardous." In contrast to the early applications on geophysical scales, the term "large" called for some qualification. It designated the scales affected by the boundary conditions (which could apply to rather small laboratory dimensions of engineering interests), while the small sub-grid scales were assumed to display more universal features: "The LES approach lies between the extremes of direct simulation, in which all fluctuations are resolved and no model is required, and the classical approach of O. Reynolds, in which only mean values are calculated and all fluctuations are modeled" (Rogallo and Moin 1984, p. 102).

Unlike ten years before, when LES ranked highest on the five-points-list of computational approaches on turbulence simulation, the LES approach was now regarded as intermediate between the (not yet accessible) resolution of all scales in DNS and the turbulence modelling by closure assumptions. For the time being, however, DNS was confined to flows at very low Reynolds numbers and therefore used only for comparative purposes. "In addition to calculating model parameters, direct simulations are also used to determine how well the forms of the SGS models represent 'exact' SGS stresses," the review from 1984 remarked about the early use of DNS as a tool for improving sub-grid scale models (Rogallo and Moin 1984, p. 110).

Tangible progress of DNS as a viable approach in its own right was first achieved in the late 1980s in studies about the onset of turbulence. Yet here, too, DNS was used in the company of other methods. With respect to the computational effort it was extremely expensive. "The LES calculations required roughly 10 h of CPU time compared with the several hundred hours for the corresponding DNS computations," a review on the simulation of the transition in wall-bounded shear flows compared the effort to compute a particular boundary-layer transition (Kleiser and Zang 1991, p. 530).

---

[9]This Week's Citation Classic. Current Contents, Number 50, 10 December 1984.

When several years later DNS became ripe for a first survey article, the reviewers left no doubt "that DNS is a research tool, and not a brute-force solution to the Navier-Stokes equations for engineering problems" (Moin and Mahesh 1998, p. 539). DNS could be used as a tool to analyse features of turbulent flows that were inaccessible otherwise. It complemented experimental investigations in the laboratory. This had already been demonstrated in studies about the structure of the turbulent boundary layer: "Access to velocity, pressure, and vorticity fields in three-dimensional space and time allowed DNS to fill in the gaps in the popular notions of boundary layer structure. In retrospect, this use of DNS represented a major change in the accepted role of computations in turbulence research" (Moin and Mahesh 1998, p. 568).

Chapter 7
Chaos and Turbulence

Abstract In the 1970s the turbulence problem became associated with the theory of nonlinear dynamical systems, in the popular parlance often labelled as "chaos theory". New concepts like "strange attractors" and "fractals" resulted in the view that turbulence may be regarded as a manifestation of "deterministic chaos". While certain cases of turbulent flows appeared amenable to dynamical systems approaches, the wealth of turbulence remained difficult to be reconciled with the new paradigm. In 1989 the dynamical systems approach became subject of controversial debate at a conference under the title "Whither Turbulence".

"Chaos: A New Paradigm"—with this headline a pioneer of the dynamical systems approach explicitly adopted the terminology of Thomas Kuhn's classic *The Structure of Scientific Revolutions* in order to highlight the novelty of this approach for analysing turbulence (Ruelle 1991, Chap. 11). The beginnings of the theory of dynamical systems may be dated back to Poincaré's celestial mechanics in the late 19th century and other mathematical studies on differential equations throughout the 20th century (Aubin and Dalmedico 2002). With regard to turbulence, however, this approach was introduced more recently—and from the very beginning with the pretence to cope with the riddles of turbulence in a fundamental way from a mathematical vantage point.

7.1 Strange Attractors

It started with a paper "On the Nature of Turbulence" published in *Communications in Mathematical Physics*, and its authors were theorists who had previously little contact with the community of turbulence researchers (Ruelle and Takens 1971). "The reception was mixed, but on the whole rather cold", David Ruelle, the main author of the paper, later recalled. "I remember the physicist C. N. Yang joking about my 'controversial ideas on turbulence'—a fair description of the situation at the time" (Ruelle 1991, p. 66). Ruelle and his collaborator, Floris Takens, took issue

M. Eckert, *The Turbulence Problem*,
SpringerBriefs in History of Science and Technology,
https://doi.org/10.1007/978-3-030-31863-5_7

with a theory put forward in the famous Landau-Lifshitz textbook on hydrodynamics which characterized the onset of turbulence as an accumulation of Fourier modes. If the number of excited modes was increased the flow would become more and more irregular, i.e. turbulent. In contrast to this view Ruelle and Takens argued that dissipative systems (like viscous fluids) behave differently. "The idea of Landau and Lifschitz must therefore be modified" (Ruelle and Takens 1971, p. 168).

In the parlance of dynamical systems the time evolution of a vector field, such as the flow velocities of a fluid, is characterized by a system of ordinary differential equations. "The time evolution of a velocity field is given by the Navier-Stokes equations

$$\frac{dv}{dt} = X_\mu(v) \tag{7.1}$$

where X_μ is a vector field over H", this is how they referred to the Navier-Stokes equations. They regarded it as "not necessary to specify further H or X_μ" because they aimed only at qualitative features (Ruelle and Takens 1971, p. 168):

> In what follows we shall investigate the nature of the solutions of (7.1), making only assumptions of a very general nature on X_μ. It will turn out that the fluid motion is expected to become chaotic when μ increases. This gives a justification for turbulence and some insight into its meaning. To study (7.1) we shall replace H by a finite-dimensional manifold and use the qualitative theory of differential equations.

In this vein they defined the "strange attractor" as the set of points to which their system approached for $t \to \infty$. In a later article for *The Mathematical Intelligencer* Ruelle resorted to the computer as a tool for visualising strange attractors: "These points correspond to the states of a chaotic system. Strange attractors are relatively abstract mathematical objects, but computers give them some life, and draw pictures of them" (Ruelle 1980, p. 126). Ten years after he had defined strange attractors in a language that appeared abhorrent to the community of flow researchers he now presented examples which made the concept more plausible. Among them was the so-called "Lorenz attractor". It had been introduced by the meteorologist Edward Lorenz in the early 1960s as a simplified mathematical model for Rayleigh-Bénard convection (describing the two-dimensional flow of a fluid layer heated from below) (Fig. 7.1). With the Lorenz system in his mind Ruelle represented (7.1) in the form

$$\frac{d}{dt}X(t) = G_\mu(X(t))$$

where μ indicated the degree of intensity by which the system was driven to chaos. In the Lorenz system the parameter r (as a measure of the temperature difference) played this role. In pipe flow and other systems with a transition from laminar to turbulent flow μ could be taken as the Reynolds number. For $\mu = 0$ the system would be stationary, $X(t) = X_0$, corresponding to a fixed point. For small μ the flow could become periodic, $X(t) = f(\omega t)$, such as regular convection rolls in the Rayleigh-Bénard flow or Kármán vortices in the wake of an obstacle. A quasiperiodic

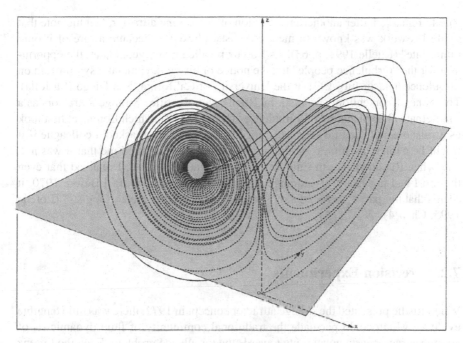

Fig. 7.1 The Lorenz system: $\frac{dx}{dt} = -\sigma x + \sigma y$, $\frac{dy}{dt} = -xy + rx - y$, $\frac{dz}{dt} = xy - bz$, as computed by Oscar Lanford and shown by David Ruelle in *The Mathematical Intelligencer*. The computation used the parameters $\sigma = 10$, $r = 28$, $b = 8/3$ and started at $\{x = 0, y = 0, z = 0\}$ for $t = 0$. The vector $\{x(t), y(t), z(t)\}$ makes loops to the right and left in an irregular manner (Ruelle 1980, Fig. 3)

flow would be characterized by $X(t) = f_k(\omega_1 t, \omega_2 t, ...\omega_k t)$. Quasiperiodic functions could display an irregular aspect, "suggestive of turbulence," Ruelle admitted, but they did not show sensitive dependence on initial conditions. Systems with a strange attractor, like the Lorenz system, critically depend on the initial conditions. They become chaotic not by accumulation of independent frequencies but by increasing the parameter μ. The difference between quasiperiodic and strange attractors shows up by Fourier analysis. The frequency spectrum of a quasiperiodic function displays discrete peaks, the strange attractor a continuum. Turbulence, with its continuous spectrum of frequencies, was prone to the strange attractor representation, but the spectrum of a quasiperiodic attractor with many frequencies is hard to distinguish from a continuous spectrum so that quasiperiodic attractors could not be ruled out a priori.

Ruelle referred to the Lorenz attractor as a prelude to the section on "Fluid Turbulence: One of the Great Unsolved Problems of Theoretical Physics". Lorenz gave "some theoretical excuse to the wellknown unreliability of weather forecasts", this is how Ruelle associated the Lorenz attractor with everyday experience (Ruelle 1980, p. 132). At the same time the affiliation of Lorenz with meteorology explained why his work from the early 1960s had escaped Ruelle's attention. "Before leaving Lorenz,"

Ruelle remarked after another presentation of the Lorenz attractor, "let me note that while his work was known to meteorologists, physicists became aware of it only rather late" (Ruelle 1991, p. 63). As Lorenz recalled many years later, the opportunity for the "turbulence people" to take notice of his work came at a symposium on turbulence held in July 1971 at the University of California, San Diego (La Jolla). The bone of contention was a talk by Ruelle with the title "Strange Attractors as a Mathematical Explanation of Turbulence" (Ruelle 1972), which Lorenz at first took as a misnomer. "The title seemed strange to me, and I even asked a colleague if it might be a mistranslation from the original French. He assured me that it was not, and when Ruelle spoke, in English at least as fluent as mine, I realized that even though I had not heard of a strange attractor, I had seen one. [...] Before 1970, it seems that my paper was cited almost exclusively by other meteorologists" (Lorenz 1995, Chap. 4).

7.2   Precision Experiments

When Ruelle presented the strange attractor concept in 1971, there was little tangible evidence which could persuade the traditional community of fluid dynamicists of strange attractors as a novel route towards the turbulence problem. Even the Lorenz system represented only a simplified model for the transition of an ordered convection to chaotic turbulence. Persuasive evidence could only come from experiments which were able to distinguish between the accumulation of instabilities according to a quasiperiodic transition to turbulence and the sudden transition from a few periodic modes to the continuum represented by a strange attractor. Ten years later the situation had changed. In his article in *The Mathematical Intelligencer* Ruelle cited a paper published in *Physics Today* by the experimentalists Harry L. Swinney and Jerry P. Gollub from the Physics Department of the City College at the City University of New York (Swinney and Gollub 1978), which led him to believe "that the onset of turbulence may well correspond to the appearance of strange attractors" (Ruelle 1980, p. 135–136).

The corroboration was due to the application of new physical tools that enabled precision measurements of the frequency spectrum in Rayleigh-Bénard convection and Couette flow. The former was the subject of experiments at the Bell Laboratories by Guenther Ahlers who focused on low-temperature investigations of hydrodynamic instabilities. In 1974 Ahlers reported in the *Physical Review Letters* about his recent experiments on Rayleigh-Bénard convection in liquid helium. "In addition to providing accurate measurements of the onset of convection and of the heat transport by the fluid under a wide range of conditions, the experiments reveal a transition to, and provide a quantitative description of, a new turbulent state", Ahlers summarized his results. "The properties of this state are described rather well by a theory developed recently by McLaughlin and Martin" (Ahlers 1974, p. 1185). John B. McLaughlin and Paul C. Martin from the Lyman Laboratory of Physics at Harvard University published in the same issue of the *Physical Review Letters* the results of numerical computations about the transition to turbulence in an improved Lorenz system. "The calculated transition seems to agree in character with the qualitative picture of the transition to turbulence suggested by Ruelle and Takens," they concluded (McLaughlin and Martin 1974, p. 1190).

The Couette flow experiments of Gollub and Swinney also involved new experimental techniques. "We have measured the time dependence of the radial component of the local fluid velocity using laser Doppler velocimetry," the experimenters remarked about their method. "The velocity is recorded digitally and then Fourier transformed to obtain velocity power spectra. The velocity spectra at different Reynolds numbers reveal the existence of several dynamical regimes which are not apparent in photographic studies or torque measurements and hence were not observed in previous studies" (Swinney et al. 1977). As Gollub and Swinney concluded in a preliminary paper: "Our observations disagree with the Landau picture of the onset of turbulence, but are perhaps consistent with proposals of Ruelle and Takens" (Gollub and Swinney 1975). Both in Rayleigh-Bénard convection and Couette flow the chaotic regime appeared "after a small number of instabilities", but there were substantial differences with regard to the number of intermediate states. "These differences can not be predicted by a universal model" (Swinney and Gollub 1978, p. 48).

Another sophisticated high-precision experiment on the transition to turbulence in Rayleigh-Bénard convection was performed at the Ecole Normale Superieure in Paris (Libchaber and Maurer 1978). In contrast to most other Rayleigh-Bénard experiments where the liquid was contained in circular chambers, the Paris group used a rectangular box. Because of its small size it was called the "Helium in a Small Box"-experiment (Libchaber and Maurer 1982). While the transition to turbulence observed by Ahlers and his collaborators at the Bell Laboratories seemed to confirm the Ruelle-Takens view, the transition in the rectangular box followed another route which had just emerged with the budding chaos theory, the so-called "Feigenbaum scenario" (discovered in 1975 and named after Mitchell Feigenbaum, a theoretical physicist from Los Alamos). The Paris experimenters observed a frequency spectrum with regularly spaced peaks which displayed the same universal scaling law as Feigenbaum's "period doubling bifurcation to chaos". Thus another part of dynamical systems theory became associated with the turbulence problem. The section in which these findings were elaborated was aptly titled "Little boxes: an exercise in dynamical system theory (Libchaber 1982, p. 1586)."

"Hence, one route to chaos in one real system may be said to be largely understood", Leo P. Kadanoff, a pioneer of the theory of phase transitions and critical phenomena, praised the agreement between experiment and theory in the case of the "little box". Nevertheless "this is only the beginning of the story—not the end", Kadanoff went on. He pointed to the transition to turbulence as observed in Ahlers' Rayleigh-Bénard experiments with a circular containment which did not follow the Feigenbaum scenario. "In fact, the period-doubling route appears to be rather rare. Are there other relatively universal routes to chaos observable in real systems? Can they also be analyzed in terms of very simple models? We do not know, but there are a large number of workers trying to find out" (Kadanoff 1983, p. 51).

7.3 Fractals

At the 1971 turbulence symposium in La Jolla, where Ruelle had introduced the concept of strange attractors, Benoit Mandelbrot, a mathematician from the IBM Thomas Watson Research Centre in New York, presented another theoretical approach beyond the established analytical tools of fluid mechanics. Mandelbrot hoped to solve the riddle of the spotty and intermittent character of fully developed turbulence by what he called the "method of self similar random multiplicative perturbations" (Mandelbrot 1972, p. 334). In a subsequent paper he discerned the "carrier of intermittent turbulence" as the subject of closer inquiry and introduced "a parameter Δ called the 'intrinsic fractional dimension' of the carrier" (Mandelbrot 1974, pp. 333). In other words: The spatial structure in which turbulence is dissipated into heat is a fractal—a notion which Mandelbrot was eager to elaborate in different contexts by the same time (Mandelbrot 1975a, 1977b, 1982). In the case of turbulence Mandelbrot extended his approach in 1975 beyond the phenomenon of intermittency and placed the geometry of turbulence in the center of his effort. "Turbulence in fluids raises a variety of interesting and practically important problems of geometry which have not, so far, received the full attention they deserve," he introduced his paper. "This failure is particularly surprising because turbulent shapes are readily visualized and therefore almost cry out for proper geometrical description" (Mandelbrot 1975b, p. 401).

In the following year the Department of Mathematics at the University of Berkeley invited Mandelbrot to deliver a lecture in a "Seminar on Turbulence: From Numerical Analysis to Strange Attractors". The aim was to discuss "the impact of dynamical systems theory", but also other "problems connected with recent theories" and "numerical work". The organizers hoped that "the seminar could give focus to serious attacks on the fundamental problem of finding a feasible model of turbulence" (Chorin et al. 1977). Mandelbrot seized this opportunity to combine "the arguments of Lorenz 1963 and Ruelle & Takens 1970" with his own approach under the label of fractals. He suggested for the strange attractor "a more positively descriptive term", such as "fractal attractor". The "dynamics approach" ought to include "fractal aspects" because "the corresponding 'worse than strange' attractor" which results from "studies à la Lorenz 1963" was undoubtedly a fractal. Mandelbrot could not yet determine its fractal dimension, but its evaluation "might help assess quantitatively rather than qualitatively to what extent natural turbulence is modeled by simplified systems of this kind" (Mandelbrot 1977a). Although Mandelbrot's suggestion to rename strange attractors as fractal attractors found few followers, "fractal aspects" soon were added to the theory of dynamical systems. The fractal dimension of strange attractors was made subject of sophisticated computational approaches and recognised as a characteristic feature of deterministic chaos (Kaplan and Yorke 1979; McGuinness 1983; Grassberger and Procaccia 1983; Farmer et al. 1983).

Turbulence served Mandelbrot as an outstanding test case not the least because it was widely known as a notorious riddle. "The study of turbulence is one of the oldest, hardest, and most frustrating chapters of physics," he opened the respective chapter in *The Fractal Geometry of Nature*. "This chapter begins with pleas for a more geometric approach to turbulence and for the use of fractals. These pleas are numerous but each is brief, because they involve suggestions with few hard results as yet" (Mandelbrot 1982, p. 97). His pleas did not fall on deaf ears. By the mid 1980s a number of "hard results" from measurements of turbulent jets, wakes and boundary layers confirmed the utility of fractals for investigations of turbulence. The theory of fractals "shed new light on some classical problems in turbulence," a study on "The fractal facets of turbulence" concluded in 1986. It helped "to quantify the seemingly complicated geometric aspects of turbulent flows, a feature that has not received its proper share of attention. The overwhelming conclusion of this work is that several aspects of turbulence can be described roughly by fractals, and that their fractal dimensions can be measured" (Sreenivasan and Meneveau 1986, p. 357) .

Mandelbrot's initial concern, the search for the "carrier of intermittent turbulence", also gave rise to further studies on fractals, usually referred to under the label of multifractals (Benzi et al. 1984; Frisch and Parisi 1985). An evaluation of experimental material on intermittency in terms of multifractals suggested "that the multifractal approach provides a useful and unifying framework for describing the scaling properties of the turbulent dissipation field... The present results exemplify the fruitful connections that can be made between the theory of nonlinear dynamical systems and turbulence (Meneveau and Sreenivasan 1987, pp. 49,72–73) ".

But fractals added little to solve old riddles. The hype with which fractals became advertised stood in sharp contrast with their utility for solving practical problems. When Hans Liepmann reviewed in 1979 "The Rise and Fall of Ideas in Turbulence" (Liepmann 1979) he ignored chaos theory and fractals, although he must have been aware of Mandelbrot's *Fractals: Form, Chance and Dimension* published two years before. "Fractals are not going to 'solve' the problem of turbulence", a reviewer qualified their impact ten years later. Nevertheless he acknowledged the role of fractal concepts for the study of numerous aspects of turbulence, such as for the analysis of data sets and scale invariance. "Many of these fall under the umbrella of fractals. It will be interesting to see what new developments evolve from numerical solutions to equations exhibiting strange-attractor behavior and from new experimental observations designed to satisfy scale-invariant conditions" (Turcotte 1988, p. 15). Another reviewer concluded three years later (Sreenivasan 1991, pp. 593–594):

> We do not truly understand why fractal and multifractal scaling holds for turbulence—at least as a reasonable approximation. Indeed, there is no satisfactory explanation of how fractals arise even in model nonlinear systems. We do not understand well the source of the fluctuations in fractal dimensions and multifractal parameters and their relation to the possible dynamical scenarios...
> Will the fractal approach survive and flourish? It is trite to say that fractals by themselves cannot solve the turbulence problem—whatever that may mean. To the extent that these are mere tools, the future depends on how intelligently and judiciously they are employed.

Despite little hope with regard to the grand riddle of turbulence, the dynamical systems approach became indispensable in some quarters of turbulence research. A study on the intermittent behaviour of a turbulent boundary layer close to a wall, for example, provided "a reasonably coherent link between low-dimensional chaotic dynamics and a realistic turbulent open flow system" (Aubry et al. 1988).

7.4 Coherent Structures

'Order and chaos' became a recurrent theme in the theory of dynamical systems. But the investigation of coherent structures in a turbulent flow had long before become part of the agenda of turbulence research in its own right. When Hans Liepmann reviewed in 1952 "Aspects of the Turbulence Problem" he called attention to "the fact that in turbulent shear flow a type of superstructure" plays a decisive role. He referred to wartime experiments on turbulent jets by Stanley Corrsin (see Sect. 4.3) and early postwar research in England by Alan Townsend on the turbulent wake of a cylinder (Townsend 1947). At the Marseille Symposium in 1961 (see Sect. 5.5) Liepmann once more portrayed "the discovery of the existence of coherent regions of nearly homogeneous vorticity fluctuations separated from the rest of the fluid by sharply defined interfaces" as one of the most important experimental findings since WW II (CNRS 1962, p. 215).

Since then the measurement of coherent structures became an ever increasing challenge for experimental turbulence investigations. Researchers from the Department of Mechanical Engineering at Stanford University discovered by a new technique of visualization coherent motions in the form of streaks which originated in the laminar sublayer of a turbulent boundary layer. "The streaks interact with the outer portions of the flow through a process of gradual 'lift-up', then sudden oscillation, bursting, and ejection. It is felt that these processes play a dominant role in the production of new turbulence and the transport of turbulence within the boundary layer on smooth walls." (Kline et al. 1967, p. 741) . At CalTech the confluence of gases under high pressure was visualized by shadowgraphs (Fig. 7.2) which revealed sequences of large coherent eddies in the turbulent mixing zone (Rebollo 1973; Brown and Roshko 1974; Roshko 1991).

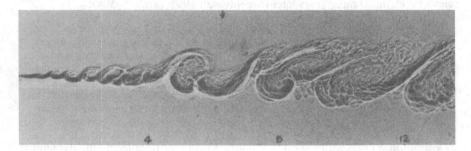

Fig. 7.2 A shadowgraph reveals coherent structures in the turbulent mixing layer of confluent jets of nitrogen and helium (Roshko 1991, Fig. 3)

Such images became icons for "A new look" on turbulence in shear flows (Roshko 1976). "Approximately twenty years have passed since the earliest observations of organized structure", a reviewer remarked in 1981, yet "a truly predictive theory" was not in sight. The review closed with another incantation of the big riddle: "Turbulence remains a major unsolved problem of classical physics" (Cantwell 1981, p. 510). Liepmann's conclusion about the rise and fall of ideas in turbulence research was similar (Liepmann 1979, p. 228):

> Clearly, the exploration of the concept of coherent structure is still on the rise. Turbulence is and will remain the most difficult problem of fluid mechanics, and past experience suggests that a subsequent fall of interest in the coherent structures is more than likely. The resulting net gain in understanding of turbulence may be less than our high expectations of today but will certainly be positive.

7.5 "Whither Turbulence"

By the 1980s leading representatives of turbulence research voiced a growing concern about diverging tendencies of their discipline. "Turbulence is rent by factionalism," the leader of the turbulence group at Cornell University, John Lumley, expressed this feeling in a flyer with which he convened a workshop titled "Whither Turbulence? Turbulence at the Crossroads" (Lumley 1990, p. 96):

> Traditional approaches in the field are under attack [...] Coherent structure people sound like *The Emperor's New Clothes* when they say that all turbulent flows consist primarily of coherent structures, in the face of visual evidence to the contrary. Dynamical systems theory people are sure that turbulence is chaos. Simulators have convinced many that we will be able to compute anything within a decade. Modeling is thus attacked as unnecessary or irrelevant because it starts with Reynolds averaging or ignores coherent structures. The card-carrying physicists dismiss everything that has been done on turbulence from Osborne Reynolds until the last decade.

The workshop, held at Cornell University in March 1989, brought to the fore how the turbulence problem was perceived a hundred years after Reynolds's pioneering work. By any accounts—textbooks, curricula, institutes, journals, conferences—turbulence in the 1980s lived up to the demands of a full-fledged discipline. Not all participants of the workshop would agree with Lumley that their discipline was at the crossroads, but there could be no doubt about the growing diversity of turbulence research during the past few decades. Liepmann had already conveyed this tendency in 1979 by alluding to a then famous cartoon about a puzzled looking Dr. Quimsey who had spent years of research to solve a problem, "but now has forgotten the question." In a similar vein, the turbulence problem appeared elusive because it escaped a straight-forward definition. "Progress in many directions has been made, indeed significant progress," Liepmann phrased its diversity. "However, the 'turbulence problem' as a whole—whatever that means—remains" (Liepmann 1979, p. 221).

At the Cornell workshop the elusiveness of the turbulence problem surfaced once more in a session on "The Utility and Drawbacks of Traditional Approaches." Liepmann's former student, Roddam Narasimha, argued that apart from few highlights like the K41-theory "the traditional approaches have mainly been a story of failures" (Lumley 1990, pp. 16–24). However, there was no clear-cut distinction between traditional and non-traditional approaches. The former was mainly identified with statistical theories, the latter with investigations of coherent structures, dynamical system theory and computational approaches (Direct and Large Eddy Simulation, Cellular Automata, Phenomenological Modelling).

Dynamical system theory, in particular, was regarded with great hopes. Its proponent (Philip Holmes) portrayed it rather as a toolbox than a single theory. The dynamical systems approach makes use of "a loose battery of theorems and methods applicable to the study of nonlinear ordinary and partial differential equations (ODE and PDE), largely developed over the past twenty or thirty years." But it could help to solve only "specific *pieces* of the turbulence problem", he cautioned against inflated expectations. The dynamical systems approach was confined to low-dimensional systems. "If one regards the 'turbulence problem' as addressing the transport of energy in the inertial (Kolmogorov) range and its subsequent dissipation, then these methods do nothing to help solve it" (Lumley 1990, pp. 196, 238). One commentator (Hassan Aref) regarded the viewpoint that chaos, and likewise turbulence, was produced in a deterministic process as the most intriguing aspect of dynamical systems. "This is a very important step (if true) in the development of our understanding of turbulence, and I suspect it would have been considered at least a qualitative solution to 'the turbulence problem' by many of the grand old men in the field" (Lumley 1990, p. 259). Another commentator (Katepalli Sreenivasan) concluded (Lumley 1990, p. 287):

> Just as initial exaggerations on the importance of the dynamical systems approach to the turbulence problem were unhealthy, so are gloomy thoughts that the approach is but a passing fad to be disparaged. The success one can have with the application of these new tools depends to some measure on how well one understands them, and on how well one can reduce sophisticated mathematical formalisms to realizable measurements.

The debates about the turbulence problem at the Cornell workshop reflect the recurrent effort of turbulence researchers to consolidate their field (Turnbull 1995)— with much more urgency than in all other turbulence symposia since Marseille 1961 (see Sect. 5.5). It went hand in hand with a concern about funding. "The fact is that funding for turbulence research is too low and we have taken an awful beating in recent years," the proponent of coherent structures research (Brian Cantwell) lamented, "the funding agencies just got tired of hearing that the solution to the turbulence problem is just around the corner" (Lumley 1990, p. 123):

> Turbulence should not be regarded as just another engineering discipline. Progress in almost any technical endeavor involving movement at sea or in the air or involving industrial processing of a fluid is nearly always limited in one way or another by the problem of turbulence. We have not been effective at getting this message across to the funding agencies and as a result contract monitors in fluid mechanics have not been able to compete effectively with their counterparts in other disciplines.

Thus the turbulence problem was given a strategic role. Should it be presented to the funding agencies as basic research or practical engineering? Perhaps the former was the better strategy. "Physicists have done this effectively for many years. Our job is more difficult in that the list of priorities must satisfy both scientific and engineering needs" (Lumley 1990, p. 124).

Chapter 8
Turbulence as a Challenge for the Historian

Abstract From an epistemological perspective, turbulence eludes the Kuhnian model of scientific revolutions. A more appropriate scheme would have to bridge the gap between the history of science on the one side and technology on the other. The historical development of turbulence in both realms suggests its analysis in terms of an evolutionary model.

When the turbulence problem is presented to broader audiences—usually in an attempt to expose the persistent challenge—it often comes about with a kind of great-mystery-rhetoric. Theodore von Kármán introduced the chapter on turbulence in his autobiography in the following way (von Kármán 1967, p. 134):

> Turbulence was, and still is, one of the great unsolved mysteries of science, and it intrigued some of the best scientific minds of the day. Arnold Sommerfeld, the noted German theoretical physicist of the 1920's, once told me, for instance, that before he died he would like to understand two phenomena—quantum mechanics and turbulence. Sommerfeld died in 1924. I believe he was somewhat nearer to an understanding of the quantum, the discovery that led to modern physics, but no closer to the meaning of turbulence.

Sommerfeld died in 1951, not in 1924. Disregarding errors with dates there is reason to doubt such recollections. Almost the same reminiscence was put in the mouth of Horace Lamb, the author of the legendary *Hydrodynamics*. Sydney Goldstein, himself famous for his textbook on *Modern Developments in Fluid Dynamics*, recalled Lamb saying at a meeting of the British Association in London in 1932 (Goldstein 1969, p. 23):

> I am an old man now, and when I die and go to Heaven there are two matters on which I hope for enlightenment. One is quantum electrodynamics, and the other is the turbulent motion of fluids. And about the former I am really rather optimistic.

Goldstein admitted that he quoted Lamb from memory and that he had heard "a similar story since repeated with other names than Lamb and other times and places." Nevertheless he concluded: "Lamb was correct on two scores. All who knew him agreed that it was Heaven that he would go to, and he was right to be more optimistic about quantum electrodynamics than turbulence" (Goldstein 1969, p. 23).

The lack of historical authenticity may be excused by the humorous tone and anecdotal character with which these utterances were made. When phrased with such great-mystery-rhetoric, however, they result in a biased historical perspective of the turbulence problem because they suggest the same research interests as in sciences that are regarded as "pure" and "fundamental science". As the preceding chapters have shown, turbulence has a long history of "applied" and "fundamental" research, if such dichotomies are at all appropriate in a field like fluid dynamics where the blurring of boundaries between science and engineering is a characteristic feature. Hugh Dryden, for example, cautioned against a sweeping use of the terms "pure", "fundamental" and "basic research" on the one hand and "applied research" on the other, "for what seems basic or fundamental to one group is regarded as applied research by another" (Dryden 1954, p. 118).

The same concern showed up at the workshop "Whither Turbulence—Turbulence at the Crossroads" (see Sect. 7.5) in discussions on the strategic role of the turbulence problem vis à vis funding agenies ("Turbulence should not be regarded as just another engineering discipline"). In a similar vein the rise of non-traditional approaches like chaos theory caused worries about "factionalism" within the community of turbulence researchers. "Turbulence is, after all, what we do for a living, and it is hard to be objective about events that affect our livelihood," John Lumley addressed this issue. "We have a situation here in which a faction in the community perceives that a condition exists in our field, and decides to take action to change the condition, while other factions of the community react to the action." Lumley alluded to Thomas Kuhn's *The Structure of Scientific Revolutions* (Kuhn 1970) where such situations are phrased in the parlance of paradigm change: "At that point the [scientific community] is divided into competing camps or parties, one seeking to defend the old institutional constellation, the other seeking to institute some new one" (Lumley 1990, p. 50).

In view of such reflections by the actors themselves it is tempting to regard the history of turbulence from the vantage point of Kuhn's scheme of scientific revolutions. Isn't the rise of non-traditional approaches like chaos theory an example of paradigm change? It is not accidental that James Gleick dedicated an entire chapter to the Kuhnian view in his popular account on *Chaos: making a new science* (Gleick 1987, Chap. 2:"Revolution") . On the long run, however, revolutions in fluid mechanics appear more gradual and affected by practical concerns which are beyond Kuhn's philosophy. The Kuhnian scheme has been thoroughly scrutinized and seldom found to fit with specific cases of scientific change without further qualifications; fluid mechanics in general, and turbulence in particular, has been ignored in this scrutiny, but it certainly would not have contradicted the conclusions from other cases (Blum et al. 2016). As Olivier Darrigol has observed for the history of hydrodynamics from the 18th to the early 20th century, the changes in this field have mostly been induced by feedback effects of applications. "They do not lead to the overthrow of the theory, yet they entail transformations of such a magnitude that the word 'application' sounds inadequate. The phenomena are not passively subjected to a rigidly established theory, but instead react upon the content and structure of the theory. They *challenge* the theory and may thus induce important adaptive transformations." Such transformations do not fit with the Kuhnian view of scientific change, "neither the smooth paradigmatic phase nor the revolutionary one" (Darrigol 2005, pp. 323–324).

A similar observation may be made for the history of turbulence. "Turbulence research lacks an agreed paradigm and is riven with controversy, yet the practitioners are able to construct a coherent field of research" (Turnbull 1995, p. 9). This conclusion resulted from an analysis of contemporary turbulence research from a sociological perspective. Instead of Kuhnian paradigm changes the continuous reconstructions or "adaptive transformations" (to quote Darrigol) appear more appropriate to account for the changing perceptions of the turbulence problem. From a more general philosophical vantage point, Stephen Toulmin had suggested earlier to compare scientific change with biological evolution. In this view "historical continuity and change can be seen as alternative results of variation and selective perpetuation, reflecting the comparative success with which different variants meet the current demands to which they are exposed" (Toulmin 1972, p. 141). In the same vein, Walter Vincenty, a historian of technology, described aeronautical engineering in terms of a variation-selection model (Vincenti 1990, Chap. 8). Its pertinence to technological innovation (Vincenti 2000) as well as scientific change suggests that turbulence with its ramifications in both technology and science is an ideal test case for an evolutionary model.

To conclude: The historian of turbulence should not crave for one or another alleged paradigm change but rather explore the mutations and adaptations of "the turbulence problem" in ever changing scientific and technological environments.

References

Agostini, L., & Bass, J. (1950). *The Theories of Turbulence*, volume NACA TM 1377.

Ahlers, G. (1974). Low-temperature studies of the rayloigh bénard instability and turbulence. *Physical Review Letters*, *33*(20), 1185–1188.

Alkemade, F. (1995). Biography. In F. T. M. Nieuwstadt & J. A. Steketee *Selected papers of J. M. Burgers* (pp. xi–cix). Dordrecht: Kluewer.

Aspray, W. (1990). *John von Neumann and the origins of modern computing*. Cambridge, MA: MIT-Press.

Aubin, D., & Dalmedico, A. (2002). Writing the history of dynamics systems and chaos: Longue durée and revolution, disciplines and cultures. *Historia Mathematica*, *29*, 1–67.

Aubry, N., Holmes, P., Lumley, J. L., & Stone, E. (1988). The dynamics of coherent structures in the wall region of a turbulent boundary layer. *Journal of Fluid Mechanics*, *192*, 115–173.

Bacon, D. L., & Reid, E. G. (1924). The resistance of spheres in wind tunnels and in air. *NACA Report*, *185*.

Bagge, E., Diebner, K., & Jay, K. (1957). *Von der Uranspaltung bis Calder Hall*. Rowohlt.

Bashe, C. J. (1982). The SSEC in historical perspective. *Annals of the History of Computing*, *4*(4), 296–312.

Batchelor, G. (1947). Kolmogoroff's theory of locally isotropic turbulence. *Proceedings of the Cambridge Philosophical Society*, *43*, 533–559.

Batchelor, G. K. (1946). Double velocity correlation function in turbulent motion. *Nature*, *158*(4024), 883–884.

Batchelor, G. K. (1948). Recent developments in turbulence research. In H. Levy (ed.), *Proceedings of the 7th International Congress for Applied Mechanics*, London.

Batchelor, G. K. (1953). *The theory of homogeneous turbulence*. Cambridge: Cambridge University Press.

Batchelor, G. K. (1996). *The life and legacy of G I. Taylor*. Cambridge: Cambridge University Press.

Battimelli, G. (1984). The mathematician and the engineer: Statistical theories of turbulence in the 20's. *Rivista di storia della scienza*, *1*, 73–94.

Battimelli, G. (1988). The early international congresses of applied mechanics. In S. Juhasz (ed.), *IUTAM. A Short History* pp. 9–13. Berlin: Springer.

Battimelli, G., & Vulpiani, A. (1982). *Kolmogorov, Heisenberg, von Weizsäcker, Onsager: un caso di scoperta simultanea* (pp. 169–175). Palermo: Atti del Terzo Congresso Nazionale di Storia della Fisica.

© The Author(s), under exclusive license to Springer Nature Switzerland AG 2019
M. Eckert, *The Turbulence Problem*,
SpringerBriefs in History of Science and Technology,
https://doi.org/10.1007/978-3-030-31863-5

Benzi, R. (2011). Lewis Fry Richardson. In P. A. Davidson, Y. Kaneda, K. Moffatt, & K. R. Sreeni-vasan (Eds.), *A voyage through turbulence* (pp. 187–208). Cambridge: Cambridge University Press.

Benzi, R., Paladin, G., Parisis, G., & Vulpiani, A. (1984). On the multifractal nature of fully developed turbulence and chaotic systems. *Journal of Physics A: General Physics, 17*(18), 3521–3531.

Betchov, R., William, O., & Criminale, J. (1967). *Stability of parallel flows.* New York, London: Academic Press.

Biezeno, C. B., & Burgers, J. M. (eds.) (1924). *Proceedings of the first International Congress for Applied Mechanics, Delft.*

Blasius, H. (1911). *Das Ähnlichkeitsgesetz bei Reibungsvorgängen. Physikalische Zeitschrift, 12,* 1175–1177.

Blasius, H. (1912). Luftwiderstand und Reynoldssche Zahl. (Das Ähnlichkeitsgesetz bei Reibungsvorgängen). *Zeitschrift für Flugtechnik und. Motorluftschiffahrt, 3*(3), 36–37.

Blasius, H. (1913). Das Ähnlichkeitsgesetz bei Reibungsvorgängen in Flüssigkeiten. *Forschungsarbeiten auf dem Gebiete des Ingenieurwesens, 131.*

Blum, A., Gavroglu, K., Joas, C., & Renn, J. (Eds.). (2016). *Shifting paradigms: Thomas S. Kuhn and the history of* (Science ed.). Berlin: Open Access.

Blumenthal, O. (1913). Zum Turbulenzproblem. *Sitzungsberichte der mathematisch-physikalischen Klasse der K. Bayerischen Akademie der Wissenschaften,* pp. 563–595.

Bodenschatz, E., & Eckert, M. (2011). Prandtl and the Göttingen school. In P. A. Davidson, Y. Kaneda, K. Moffatt, & K. R. Sreenivasan, (Eds.), *A Voyage Through Turbulence* (pp. 40–100). Cambridge: Cambridge University Press.

Boussinesq, J. (1897). *Théorie de l'écoulement tourbillonnant et tumultueux des liquides dans les lits rectilignes a grande section.* Paris: Gauthier-Villars.

Bradshaw, P. (1972). The understanding and prediction of turbulent flow. *Aeronautical Journal, 76*:403–418. Presented at the Sixth Reynolds-Prandtl Lecture, given in Munich on 14th April 1972.

Breford, & Möller, E. (1943). Messungen am Originalflügel des Baumusters P-51 'Mustang'. Deutsche Luftfahrtforschung, Forschungsbericht 1724-2. *ZWB.*

Brennan, J. F. (1971). *The IBM Watson laboratory at Columbia University: A history.* Armonk, NY: IBM.

Brown, G. L., & Roshko, A. (1974). On density effects and large structure in turbulent mixing layers. *Journal of Fluid Mechanics, 64*(4), 775–816.

Brown, W. B. (1959). Numerical calculation of the stability of cross flow profiles in laminar boundary layers on a rotating disc and on a swept back wing and an exact calculation of the stability of the Blasius velocity profile. *Northrop Aircraft Inc., Report NAI 59-5.*

Burgers, J. M. (1925). The motion of a fluid in the boundary layer along a plane smooth surface. In C. B. Biezeno & J. M. Burgers (Eds.), *Proceedings of the First International Congress for Applied Mechanics Delft 1924* (pp. 113–128).

Burgers, J. M. (1939). Mathematical examples illustrating relations occurring in the theory of turbulent fluid motion. *Verhandelingen der Koninklijke Nederlandse Academie van Wetenschappen, Afdeeling Natuurkunde, XVI, I*(2), 1–53.

Burgers, J. M. (1940a). Application of a model system to illustrate some points of the statistical theory of free turbulence. *Proceedings of the Academy of Science Amsterdam, 43,* 2–12.

Burgers, J. M. (1940b). On the application of statistical mechanics to the theory of turbulent fluid motion. A hypothesis which can serve as a basis for a statistical treatment of some mathematical model systems. In *Proceedings of the Academy of Science Amsterdam, 43,* 936–945, 1153–1159.

Burgers, J. M. (1948). A mathematical model illustrating the theory of turbulence. *Advances in Applied Mechanics, 1,* 171–199.

Burgers, J. M. (1975). Some memories of early work in fluid mechanics at the Technical University of Delft. *Annual Review of Fluid Mechanics, 7,* 1–11.

Bußmann, K. (1942). Experimentelle und theoretische Untersuchungen an Laminarprofilen. Deutsche Luftfahrtforschung, Forschungsbericht 1709-1. *ZWB.*

Bußmann, K. (1943). Messungen am Laminarprofil P-51 'Mustang'. Deutsche Luftfahrtforschung, Forschungsbericht 1724-1. *ZWB.*

Cantwell, B. J. (1981). Organized motion in turbulent flow. *Annual Review of Fluid Mechanics, 13,* 457–515.

Cebeci, T., & Smith, A. M. O. (1974). *Analysis of turbulent boundary layers.* New York: Academic Press.

Chentsov, N. N. (1990). The unfathomable influence of kolmogorov. *The Annals of Statistics, 1*(3), 987–998.

Chorin, A., Marsden, J., & Smale, S. (1977). *Turbulence Seminar. Berkeley 1976/77.* Berlin, Heidelberg, New York: Springer.

CNRS. (1962). *La Mécanique de la Turbulence.* Paris: CNRS.

Corrsin, S. (1943). Investigation of Flow in an Axially Symmetrical Heated Jet of Air. *NACA Advance Confidential Report,* 3L23.

Corrsin, S. (1944). Investigation of the Behavior of Parallel Two-Dimensional Air Jets. *NACA Advance Confidential Report,* 4H24.

Corrsin, S. (1961). Turbulent flow. *American Scientist, 49*(3), 300–325.

Darrigol, O. (2002). Turbulence in 19th century hydrodynamics. *Historical Studies in the Physical Sciences, 32*(2), 207–262.

Darrigol, O. (2005). *Worlds of flow.* Oxford: Oxford University Press.

Deardorff, J. (1970). A numerical study of three-dimensional turbulent channel flow at large Reynolds numbers. *Journal of Fluid Mechanics, 41*(2), 453–480.

Deardorff, J. W. (1973). The use of subgrid transport equations in a three-dimensional model of atmospheric turbulence. *Journal of Fluids Engineering, 95,* 429–438.

Doenhoff, A. E. V., & Tetervin, N. (1943). Determination of general relations for the behavior of turbulent boundary layers. *NACA Advance Confidential Report,* No. 3G13.

Doetsch, H. (1943). Versuche am Tragflügel des North-American Mustang. Teil 1. Deutsche Luftfahrtforschung, Forschungsbericht 1712. *ZWB.*

Drazin, P. G., & Reid, W. H. (2003). *Hydrodynamic stability.* Cambridge University Press, Cambridge, second edition edition.

Dryden, H. L. (1937). The theory of isotropic turbulence. *Journal of the Aeronautical Sciences, 4*(7), 273–280.

Dryden, H. L. (1939). Turbulence investigations at the national bureau of standards. In *Proceedings of the Fifth International Congress on Applied Mechanics, Cambridge Mass.,* J. P. Den Hartog & H. Peters (eds), Wiley, New York, pp. 362–368.

Dryden, H. L. (1947). The international congress for applied mechanics. *Science, 105*(2720), 167–169. 14 February 1947.

Dryden, H. L. (1951). The turbulence problem today. In *Proceedings of the Midwestern Conference on Fluid Dynamics: First Conference,* May 12–13. (1950). University of Illinois, pp. 1–20. Ann Arbor, Michigan: Edwards.

Dryden, H. L. (1954). A half century of aeronautical research. *Proceedings of the American Philosophical Society, 9,* 115–120.

Dryden, H. L. (1955). Fifty years of boundary-layer theory and experiment. *Science, 121*(3142), 375–380.

Dryden, H. L., & Heald, R. H. (1926). Investigation of turbulence in wind tunnels by a study of the flow about cylinders. *NACA Report,* 231.

Dryden, H. L. & Kuethe, A. M. (1929a). Effect of turbulence in wind tunnels measurements. *NACA Report,* 342.

Dryden, H. L., & Kuethe, A. M. (1929b). The measurement of fluctuations of air speed by the hot-wire anemometer. *NACA Report,* 320.

Dryden, H. L., Schubauer, G. B, Jr., & W. C. M., & Skramstad, H. K., (1936). Measurements of intensity and scale of wind-tunnel turbulence and the relation to critical Reynolds number of spheres. *NACA Report, 581,* 109–140.

Durbin, P. A. (2017). Some recent developments in turbulence closure modeling. *Annual Review of Fluid Mechanics, 50,* 77–103.

Ecke, R. (2005). The turbulence problem. An experimentalist's perspective. *Los Alamos Science, 29,* 124–141.

Eckert, M. (2006). *The dawn of fluid dynamics.* Weinheim: Wiley-VCH.

Eckert, M. (2010). The troublesome birth of hydrodynamic stability theory: Sommerfeld and the turbulence problem. *European Physical Journal, History, 35*(1), 29–51.

Eckert, M. (2017). *Ludwig Prandtl-Strömungsforscher und Wissenschaftsmanager.* Berlin, Heidelberg: Springer.

Eckert, M. (2018). Turbulence Research in the 1920s and 1930s between mathematics, physics, and engineering. *Science in Context, 31*(3), 381–404.

Eckert, M. (2019a). *Ludwig Prandtl: A Life for Fluid Mechanics and Aeronautical Research.* Translated by David A. Tigwell. Heidelberg: Springer International Publishing.

Eckert, M. (2019b). *Strömungsmechanik zwischen Mathematik und Ingenieurwissenschaft: Felix Kleins Hydrodynamikseminar 1907–08.* Hamburg: Hamburg University Press.

Edwards, P. N. (2010). *A vast machine: Computer models, climate data, and the politics of global warming.* Cambridge, MA: MIT Press.

Emmons, H. W. (1970). Critique of numerical modeling of fluid-mechanical phenomena. *Annual Reviews of Fluid Mechanics, 2,* 15–36.

Epple, M. (2002). Rechnen, Messen, Führen. Kriegsforschung am KWI für Strömungsforschung 1937-1945. In Helmut Maier (Hg.), *Rüstungsforschung im Nationalsozialismus. Organisation, Mobilisierung und Entgrenzung der Technikwissenschaften* (pp. 305–356). Göttingen: Wallstein.

Eyink, G. L., & Sreenivasan, K. R. (2006). Onsager and the theory of hydrodynamic turbulence. *Reviews of Modern Physics, 78*(1), 87–135.

Falkovich, G. (2011). The Russian school. P. A. Davidson, Y. Kaneda, K. Moffatt, & K. R. Sreenivasan, (Eds.), *A voyage through turbulence* (pp. 209–237). Cambridge: Cambridge University Press.

Farge, M., Moffatt, H. K., & Schneider, K., (eds.) (2011). *Fundamental problems of turbulence: Fifty years after the turbulence colloquium Marseille of 1961.* edp Sciences.

Farmer, J. D., Ott, E., & Yorke, J. A. (1983). The dimension of chaotic attractors. *Physica, 7D,* 153–180.

Ferziger, J. H. (1977). Large eddy numerical simulations of turbulent flows. *AIAA Journal, 15*(9), 1261–1267.

Forchheimer, P. (1905). Hydraulik. *Enzyklopädie der mathematischen Wissenschaften, IV*(20), 324–472.

Fox, D. G., & Lilly, D. K. (1972). Numerical simulation of turbulent flows. *Reviews of Geophysics and Space Physics, 10*(1), 51–72.

Friedlander, S. K., & Topper, L. (Eds.). (1961). *Turbulence: Classic papers on statistical theory.* New York and London: Interscience Publishers.

Frisch, U. (1995). *Turbulence. The Legacy of A. N. Kolmogorov:* Cambridge University Press, Cambridge.

Frisch, U., & Bec, J. (2001). Burgulence. In M. Lesieur, A. Yaglom, & F. David (Eds.), *New trends in turbulence Turbulence: nouveaux aspects. Les Houches - Ecole d'Ete de Physique Theorique* (vol 74, pp. 341–383). Berlin, Heidelberg: Springer.

Frisch, U., & Parisi, G. (1985). Fully developed turbulence and interniittency. In M. Ghil, R. Benzi & G. Parisi (Eds.) *Turbulence and predictability in geophysical fluid dynamics and climate dynamics, Varenna, 1983, Proceedings of the International School of Physic Enrico Fermi* (pp. 71–88). North-Holland.

Fritsch, W. (1928). Der Einfluß der Wandrauhigkeit auf die turbulente Geschwindigkeitsverteilung in Rinnen. *Zeitschrift für Angewandte Mathematik und Mechanik (ZAMM), 8,* 199–216.

Föppl, O. (1911). Windkräfte an ebenen und gewölbten Platten. *Jahrbuch der Motorluftschiff-Studiengesellschaft, 4,* 51–119.

Galperin, B., & Orszag, S. A. (Eds.). (1993). *Large-eddy simulation of complex engineering and geophysical flows.* Cambridge: Cambridge University Press.

Gericke, H. (1972). *50 Jahre GAMM.* Berlin: Springer.

Gleick, J. (1987). *Chaos: Making a new science.* New York: Vicking.

Goering, H. (ed.) (1958). *Sammelband zur statistischen Theorie der Turbulenz, mit Beiträgen von A. N. Kolmogorov, A. M. Obuchow, A. M. Jaglom, A. S. Monin.* Akademie-Verlag, Berlin (East).

Goldstein, S. (1969). Fluid mechanics in the first half of this century. *Annual Reviews of Fluid Mechanics, 1,* 1–29.

Gollub, J. P., & Swinney, H. L. (1975). Onset of turbulence in a rotating fluid. *Physical Review Letters, 35*(14), 927–930.

Grant, H. L., Stewart, R. W., & Moilliet, A. (1962). Turbulence spectra from a tidal channel. *Journal of Fluid Mechanics, 12,* 241–268.

Grassberger, P., & Procaccia, I. (1983). Measuring the strangeness of strange attractors. *Physica, 9D,* 189–208.

Grinstein, F. F., Margolin, L. G., & Rider, W. J. (Eds.). (2007). *Implicit large eddy simulation: Computing turbulent fluid dynamics.* Cambridge: Cambridge University Press.

Görtler, H. (1948). Turbulenz. *Naturforschung und Medizin in Deutschland 1939-1946. Für Deutschland bestimmte Ausgabe der FIAT Review of German Science, 5*(III), 75–100.

Hagen, G. (1854). Über den Einfluss der Temperatur auf die Bewegung des Wassers in Röhren. *Mathematische Abhandlungen der königlichen Akademie der Wissenschaften zu Berlin,* 17–98.

Hager, W. H. (1994). Die historische Entwicklung der Fliessformel. *Schweizer Ingenieur und Architekt, 112*(9), 123–133.

Hager, W. H. (2003). Blasius: A life in research and education. *Experiments in Fluids, 34,* 566–571.

Hahn, H., Herglotz, G., & Schwarzschild, K. (1904). Über das Strömen des Wassers in Röhren und Kanälen. *Zeitschrift für Mathematik und Physik, 51,* 411–426.

Hamel, G. (1911). *Zum Turbulenzproblem* (pp. 261–270). Mathematisch-Physikalische Klasse: Göttinger Nachrichten der Kgl. Gesellschaft der Wissenschaften.

Hansen, M. (1928). Die Geschwindigkeitsverteilung in der Grenzschicht an einer eingetauchten Platte. *Zeitschrift für Angewandte Mathematik und Mechanik (ZAMM), 8,* 185–199.

Harlow, F. H., & Fromm, J. E. (1963). Numerical solution of the problem of vortex street development. *The Physics of Fluids, 6*(7), 975–982.

Harper, K. C. (2008). *Weather by the numbers: The genesis of modern meteorology.* Cambridge, Massachusetts: The MIT Press.

Hartog, J. P. D., & Peters, H., (eds.) (1939). *Proceedings of the Fifth International Congress for Applied Mechanics.* Wiley, New York. Held at Harvard University and the Massachusetts Institute of Technology, Cambridge, Massachusetts, September 12–16, 1938.

Haupt, O. (1912). Über die Entwicklung einer willkürlichen Funktion nach den Eigenfunktionen des Turbulenzproblems. *Sitzungsberichte der mathematisch-physikalischen Klasse der K. Bayerischen Akademie der Wissenschaften,* pp. 289–301.

Heisenberg, W. (1922). Nichtlaminare Lösungen der Differentialgleichungen für reibende Flüssigkeiten. In T. von Kármán & T. Leci-Civita (Eds.), *Vorträge aus dem Gebiete der Hydro- und Aerodynamik (Innsbruck 1922)* (pp. 139–142). Berlin: Springer.

Heisenberg, W. (1924). Über Stabilität und Turbulenz von Flüssigkeitsströmen. *Annalen der Physik, 74,* 577–627.

Heisenberg, W. (1948a). Bemerkungen zum Turbulenzproblem. *Zeitschrift für Naturforschung, 3a,* 434–437.

Heisenberg, W. (1948b). Zur statistischen Theorie der Turbulenz. *Zeitschrift für Physik, 124,* 628–657.

Heisenberg, W. (1969). Significance of Sommerfeld's Work Today. In F. Bopp, H. Kleinpoppen (eds.) *Physics of the one- and two-electron atom. North-Holland Publication Company,* pp. 44–52. Proceedings of the Arnold Sommerfeld Centennial Memorial Meeting and of the International Symposium on the Physics of the One- and Two-Electron Atoms. Munich, 10–14 September 1968.

Hinze, J. O. (1959). *Turbulence an introduction to its mechanism and theory.* New York: McGraw-Hill.

Holstein, H. (1947). Laminarhaltung durch Formgebung. *Monographien über Fortschritte der deutschen Luftfahrtforschung (seit 1939)*, herausgegeben von A. Betz. Band B (Grenzschichten, redigiert von W. Tollmien), Kap. 4.1, Göttingen, DLR.

Hopf, L. (1910). *Hydrodynamische Untersuchungen: Turbulenz bei einem Flusse. Über Schiffswellen. Inaugural-Dissertation.* Leipzig: Barth.

Hopf, L. (1914). Der Verlauf kleiner Schwingungen auf einer Strömung reibender Flüssigkeit. *Annalen der Physik, 44*, 1–60.

IUTAM and IAU, editors (1949). *Problems of cosmical aerodynamics. Proceedings of the Symposium on the Motion of Gaseous Masses of Cosmical Dimensions held at Paris, August 16–19, Dayton, OH.* Central Air Documents Office (Army-Navy-Air Force).

Kadanoff, L. P. (1983). Roads to chaos. *Physics Today, 36*(12), 46–53.

Kanak, K. M. (2004). Douglas K. Lilly: a biography. In E. Fedorovich, R. Rotunno, & B. Stevens (Eds.), *Atmospheric Turbulence and Mesoscale Meteorology: Scientific Research Inspired by Doug Lilly* (pp. 1–14). Cambridge: Cambridge University Press.

Kaplan, J., & Yorke, J. (1979). Chaotic behavior of multidimensional difference equations. In H. O. Peitgen, H. O. Walther (eds.) *Functional Differential Equations and the Approximation of Fixed Points. Lecture Notes in Mathematics*, vol. 730. Berlin: Springer, pp. 204–227.

Kevles, D. J. (1971). Into hostile political camps: The reorganization of international science in world war I. *Isis, 62*(1), 47–60.

Klebanoff, P. S., Tidstrom, K. D., & Sargent, L. H. (1962). The three-dimensional nature of boundary layer instability. *Journal of Fluid Mechanics, 12*, 1–34.

Kleiser, L., & Zang, T. A. (1991). Numerical simulation of transition in wall-bounded shear flows. *Annual Review of Fluid Mechanics, 23*, 495–537.

Kline, S. J., Moffatt, H. K., & Morkovin, M. V. (1969). Report on the AFOSR-IFP-Stanford conference on computation of turbulent boundary layers. *Journal of Fluid Mechanics, 36*(3), 481–484.

Kline, S. J., Reynolds, W. C., Schraub, F. A., & Runstadler, P. W. (1967). The structure of turbulent boundary layers. *Journal of Fluid Mechanics, 30*(4), 741–773.

Kluwick, A. (1996). Comments on the paper 'On the flow of water through ducts and channels' by H. Hahn, G. Herglotz and K. Schwarzschild. L. Schmetterer, K. Sigmund (eds.) *Hans Hahn. Gesammelte Abhandlungen, Collected Works.* New York: Springer, 2:513–517.

Kolmogorov, A. N. (1991a). Dissipation of energy in the locally isotropic turbulence. *Proceedings of the Royal Society London A, 434*, 15–17. First published in Russian in Dokl. Akad. Nauk SSSR (1941), 32(1).

Kolmogorov, A. N. (1991b). The local structure of turbulence in incompressible viscous fluid for very large Reynolds numbers. *Proceedings of the Royal Society London A, 434*, 9–13. First published in Russian in Dokl. Akad. Nauk SSSR (1941) 30(4).

Kuhn, T. S. (1970). *The structure of scientific revolutions.* Chicago: University of Chicago Press.

Kurtz, E. F., & Crandall, S. H. (1962). Computer-aided analysis of hydrodynamic stability. *Studies in Applied Mathematics, 41*(4), 264–279.

Landau, L. D. (1944). On the problem of turbulence. In D. Ter Haar (ed.) *Collected Papers of L. D. Landau.* Oxford: Pergamon Press, 1965, pp. 387–391.

Launder, B., & Jackson, D. (2011). Osborne Reynolds: A turbulent life. In P. A. Davidson, Y. Kaneda, K. Moffatt, K. R. Sreenivasan, (eds.) *A Voyage Through Turbulence (Cambridge: Cambridge University Press*, pp. 1–39.

Launder, B. E., Reece, G. J., & Rodi, W. (1975). Progress in the development of a Reynolds-stress turbulence closure. *Journal of Fluid Mechanics, 68*(3), 537–566.

Launder, B. E., & Spalding, D. B. (1974). The numerical computation of turbulent flows. *Computer Methods in Applied Mechanics and Engineering, 3*(2), 269–289.

Leonard, A., & (1974). Energy cascade in large-eddy simulations of turbulent fluid flows. Turbulent diffusion in environmental pollution. In Proceedings of a Symposium held at Charlottesville, Virginia, April 8–14,. (1973). *Volume 18A* (pp. 237–248). New York: Academic Press Inc.

Leonard, A., & Peters, N. (2011). Theodore von Kármán. In P. A. Davidson, Y. Kaneda, K. Moffatt & K. R. Sreenivasan (Eds.), *A voyage through turbulence* (pp. 101–126). Cambridge: Cambridge University Press.

LePage, W. L., & Nichols, J. T. (1924). The effect of wind tunnel turbulence upon the forces measured on models. *NACA Technical Notes, 191*, 1–10.

Lesieur, M., & Métais, O. (1996). New trends in large-eddy simulations of turbulence. *Annual Review of Fluid Mechanics, 28*, 45–82.

Libchaber, A. (1982). Convection and Turbulence in liquid helium I. *Physica, 109*(110B), 1583–1589.

Libchaber, A., & Maurer, J. (1978). Local probe in a Rayleigh-Bénard experiment in liquid helium. *Journal de Physique Lettres, 39*(21), 369–372.

Libchaber, A., & Maurer, J. (1982). A Rayleigh-Bénard Experiment: Helium in a Small Box. In T. Riste (Ed.) *Nonlinear Phenomena at Phase Transitions and Instabilities. NATO Advanced Study Institutes Series* (vol. 77, pp. 259–286). Boston, MA: Springer.

Liepmann, H. (1943). Investigations on laminar boundary layer stability and transition on curved boundaries. *NACA Advance Confidential Report*, (ACR 3H30).

Liepmann, H. W. (1952). Aspects of the Turbulence Problem. *Zeitschrift für angewandte Mathematik und Physik (ZAMP), 3*(6), 321–342, 407–426.

Liepmann, H. W. (1979). The rise and fall of ideas in turbulence. *American Scientist, 67*(2), 221–228.

Lighthill, M. J. (1956). Reviews: The structure of turbulent shear flow. *Journal of Fluid Mechanics, 1*(5), 554–560.

Lilienthal-Gesellschaft. (1940). Über die laminare und turbulente Reibungsschicht. Bericht S-10: Preisausschreiben 1940, Flugzeugbau. *ZWB*.

Lilienthal-Gesellschaft. (1941). Ausschuß Allgemeine Strömungsforschung, Bericht über die Sitzung Grenzschichtfragen am 28. und 29. Oktober 1941 in Göttingen. Bericht 141. *ZWB*.

Lilly, D. K. (1967). The representation of small scale turbulence in numerical simulation experiments. In J. Thomas (Ed.), *IBM scientific computing symposium on environmental sciences* (pp. 195–210). Heights: Watson Research Center, Yorktown.

Lin, C. -C. (1944). *On the Development of Turbulence*. Ph.D. thesis, California Institute of Technology, Pasadena, California.

Lin, C.-C. (1955). *The theory of hydrodynamic stability*. Cambridge: Cambridge University Press.

Livi, R., & Vulpiani, A. (Eds.). (2003). *The kolmogorov legacy in physics*. Berlin, Heidelberg: Springer.

Lorentz, H. A. (1897). Over den weerstand dien een vloeistofstroom in eene cilindrische buis ondervindt. *Versl. K. Akad. Wet. Amsterdam, 6*, 28–49.

Lorentz, H. A. (1907). Über die Entstehung turbulenter Flüssigkeitsbewegungen und über den Einfluss dieser Bewegungen bei der Strömung durch Röhren. *Hendrik Antoon Lorentz: Abhandlungen über theoretische Physik* (Leipzig: Teubner). *1*, 43–71.

Lorenz, E. N. (1995). *The essence of chaos*. UCL Press.

Love, A. E. H. (1901a). Hydrodynamik. I. Physikalische Grundlegung. *Enzyklopädie der mathematischen Wissenschaften, IV*(15), 48–83.

Love, A. E. H. (1901b). Hydrodynamik. II. Theoretische Ausführungen. *Enzyklopädie der mathematischen Wissenschaften, IV*(16), 84–147.

Lumley, J. L. (Eds.) (1990). *Whither Turbulence? Turbulence at the Crossroads. Proceedings of a Workshop Held at Cornell University, Ithaca, NY, March 22–24, 1989*, volume 357 of Lecture Notes in Physics. Berlin, Heidelberg: Springer.

Mack, L. M. (1960). *Numerical calculation of the stability of the compressible, laminar boundary layer*, pp. 20–122. Jet Propulsion Laboratory Report No: California Institute of Technology.

Mandelbrot, B. B. (1972). Possible refinement of the lognormal hypothesis concerning the distribution of energy dissipation in intermittent turbulence. In M. Rosenblatt & C. van Atta (eds.) *Statistical Models and Turbulence. Proceedings of a Symposium held at the University of Califor-*

nia, San Diego (La Jolla) July 15–21, 1971. Berlin, Heidelberg: Springer, 12:333–351. Lecture Notes in Physics.

Mandelbrot, B. B. (1974). Intermittent turbulence in self-similar cascades: Divergence of high moments and dimension of the carrier. *Journal of Fluid Mechanics, 62*(2), 331–358.

Mandelbrot, B. B. (1975a). *Les objets fractals: Form*. Flammarion, Paris: Hasard et Dimension.

Mandelbrot, B. B. (1975b). On the geometry of homogeneous turbulence, with stress on the fractal dimension of the iso-surfaces of scalars. *Journal of Fluid Mechanics, 72*(2), 401–416.

Mandelbrot, B. B. . . . (1977a). Fractals and turbulence: Attractors and dispersion. In P. Bernard, T. Ratiu (eds) *Turbulence Seminar. Lecture Notes in Mathematics* (vol 615, pp. 83–93). Berlin, Heidelberg: Springer.

Mandelbrot, B. B. (1977b). *Fractals: Form, Chance and Dimension*. New York: W. H. Freeman. translated from the French original published in 1975 under the title *Les Objets Fractals: Forme, Hasard Et Dimension*.

Mandelbrot, B. B. (1982). *The fractal geometry of nature*. New York: W. H. Freeman.

Manegold, K.-H. (1970). *Universität, Technische Hochschule und Industrie: Ein Beitrag zur Emanzipation der Technik im 19*. Duncker und Humblot, Berlin: Jahrhundert unter besonderer Berücksichtigung der Bestrebungen Felix Kleins.

Marusic, I., & Nickels, T. B. (2011). A. A. Townsend. In P. A. Davidson, Y. Kaneda, K. Moffatt & K. R. Sreenivasan (Eds.), *A voyage through turbulence* (pp. 305–328). Cambridge: Cambridge University Press.

McGuinness, M. J. (1983). The fractal dimension of the Lorenz attractor. *Physics Letters, 99A*(1), 5–9.

McLaughlin, J. B., & Martin, P. C. (1974). Transition to turbulence of a statically stressed fluid. *Physical Review Letters, 33*(20), 1189–1192.

Mellor, G. L., & Herring, H. J. (1972). A survey of the mean turbulent field closure models. *AIAA Journal, 11*(5), 590–599.

Meneveau, C., & Katz, J. (2000). Scale-invariance and turbulence models for large-eddy simulation. *Annual Review of Fluid Mechanics, 32*, 1–32.

Meneveau, C., & Sreenivasan, K. R. (1987). The Multifractal Spectrum of the Dissipation Field in Turbulent Flows. *Nuclear Physics B (Proc. Suppl.), 2*, 49–76.

Mock, W. C. J., & Dryden, H. L. (1932). Improved apparatus for the measurement of fluctuations of air speed in turbulent flow. *NACA Report, 448*, 131–154.

Moffatt, H. K. (2002). G. K. Batchelor and the homogenization of turbulence. *Annual Reviews of Fluid Mechanics, 34*, 19–35.

Moffatt, H. K. (2011). Georg batchelor: the post-war renaissance of research in turbulence. In P. A. Davidson, Y. Kaneda, K. Moffatt, & K. R. Sreenivasan (Eds.), *A Voyage Through Turbulence* (pp. 276–304). Cambridge: Cambridge University Press.

Moffatt, H. K. (2012). Homogeneous turbulence: An introductory review. *Journal of Turbulence, 13*(1), N39.

Moin, P., & Mahesh, K. (1998). Direct numerical simulation: A tool in turbulence research. *Annual Review of Fluid Mechanics, 30*, 539–578.

Morkovin, M. V. (1969). On the many faces of transition. In C. Sinclair Wells (ed.): *Viscous Drag Reduction*. New York: Plenum Press, pp. 1–31. *Proceedings of the Symposium on Viscous Drag Reduction held at the LTV Research Center*, Dallas, Texas, September 24 and 25, 1968.

Morkovin, M. V., & Reshotko, E. (1989). Dialogue on progress and issues in stability and transition research. In D. Arnal & R. Michel (eds.) *Laminar-Turbulent Transition. IUTAM Symposium Toulouse/France*, September 11 -15, 1989. Springer: Berlin u. a., pp. 3–30.

Munk, M. (1917). Bericht über Luftwiderstandsmessungen von Streben. Mitteilung 1 der Göttinger Modell-Versuchsanstalt für Aerodynamik. *Technische Berichte. Herausgegeben von der Flugzeugmeisterei der Inspektion der Fliegertruppen. Heft Nr. 4 (1. Juni 1917)*, pages 85–96, Tafel XXXX–LXIII.

Nikuradse, J. (1926). Untersuchungen über die Geschwindigkeitsverteilung in turbulenten Strömungen. *Forschungsarbeiten auf dem Gebiete des Ingenieurwesens*, 281.

Nikuradse, J. (1930). Über turbulente Wasserströmungen in geraden Rohren bei sehr grossen Reynoldsschen Zahlen. *Vorträge aus dem Gebiete der Aerodynamik und verwandter Gebiete (Aachen 1929). Edited by A. Gilles, L. Hopf and Th. v. Kármán* (pp. 63–69). Berlin: Springer.

Noether, F. (1913). Über die Entstehung einer turbulenten Flüssigkeitsbewegung. *Sitzungsberichte der mathematisch-physikalischen Klasse der K. Bayerischen Akademie der Wissenschaften*, pp. 309–329.

Noether, F. (1914a). Zur Theorie der Turbulenz. *Jahresbericht der Deutschen Mathematiker-Vereinigung, 23*, 138–144.

Noether, F. (1914b). Über den Gültigkeitsbereich der Stokesschen Widerstandsformel. *Zeitschrift für Mathematik und Physik, 63*, 1–39.

Noether, F. (1921). Das Turbulenzproblem. *Zeitschrift für Angewandte Mathematik und Mechanik (ZAMM)*, pp. 125–138, 218–219.

Noether, F. (1931). Zur statistischen Deutung der Kármánschen Ähnlichkeitshypothese in der Turbulenztheorie. *Zeitschrift für Angewandte Mathematik und Mechanik (ZAMM), 11*, 224–231.

Onsager, L. (1945). The distribution of energy in turbulence. *Physical Review, 68*, 286.

Orr, W. M. F. (1907). The stability or instability of the steady motions of a perfect liquid and a viscous liquid. *Proceedings of the Royal Irish Academy, A, 27*, 9–138.

Pekeris, C. L. (1936). On the stability problem in hydrodynamics. *Proceedings of the Cambridge Philosophical Society, 32*, 55–66.

Pekeris, C. L. (1938). On the stability problem in hydrodynamics II. *Journal of the Aeronautical Sciences, 5*(6), 237–240.

Pope, S. D. (2000). *Turbulent flows*. Cambridge. Cambridge University Press.

Prandtl, L. (1914). *Der Luftwiderstand von Kugeln* (pp. 177–190). Mathematisch-physikalische Klasse: Nachrichten der Gesellschaft der Wissenschaften zu Göttingen.

Prandtl, L. (1921). Bemerkungen über die Entstehung der Turbulenz. *Zeitschrift für Angewandte Mathematik und Mechanik (ZAMM), 1*, 431–436.

Prandtl, L. (1922). Bemerkungen über die Entstehung der Turbulenz. *Physikalische Zeitschrift, 23*, 19–25.

Prandtl, L. (1925). Bericht über Untersuchungen zur ausgebildeten Turbulenz. *ZAMM, 5*, 136–139.

Prandtl, L. (1927). *Über die ausgebildete Turbulenz* (pp. 62–75). Verhandlungen des II. Internationalen Kongresses für Technische Mechanik. Zürich: Füssli.

Prandtl, L. (1930). Vortrag in Tokyo. *Journal of the Aeronautical Research Institute, Tokyo, Imperial University, 5*(65), 12–24.

Prandtl, L. (1932). Zur turbulenten Strömung in Rohren und längs Platten. *Ergebnisse der Aerodynamischen Versuchsanstalt zu Göttingen, 4*, 18–29.

Prandtl, L. (1953). Turbulenz. Naturforschung und Medizin in Deutschland 1939–1946. *Für Deutschland bestimmte Ausgabe der FIAT Review of German Science, 11*, 55–78.

Prandtl, L., & Reichardt, H. (1934). Einfluss von Wärmeschichtung auf Eigenschaften einer turbulenten Strömung. *Deutsche Forschung, 15*, 110–121.

Prandtl, L., & Wieghardt, K. (1945). *Über ein neues Formelsystem für die ausgebildete Turbulenz* (pp. 6–19). Nachrichten der Akademie der Wissenschaften zu Göttingen: Mathematisch-physikalische Klasse.

Rayleigh, L. (1893). On the flow of viscous lliquid, especially in two dimensions. *Philosophical Magazine*, 36:354–372. Reprinted in Rayleigh's Collected Papers, vol. IV, pp. 78–93.

Rebollo, M. R. (1973). *Analytical and Experimental Investigation of a Turbulent Mixing Layer of Different Gases in a Pressure Gradient*. Ph.D. thesis, California Institute of Technology, Pasadena, CA.

Reichardt, H. (1940). Die Wärmeübertragung in turbulenten Reibungsschichten. *Zeitschrift für Angewandte Mathematik und Mechanik (ZAMM), 20*(6), 297–328.

Reid, E. G. (1925). Standardization tests of N.A.C.A. No. 1 wind tunnel. *NACA Report*, 195.

Reynolds, O. (1883). An experimental investigation of the circumstances which determine whether the motion of water in parallel channels shall be direct or sinuous and of the law of resistance in parallel channels. *Philosophical Transactions of the Royal Society, 174*, 935–982.

Reynolds, O. (1884). On the two manners of motion of water. *Proceedings of the Royal Institution of Great Britain, 11*, 44–52.

Reynolds, O. (1895). On the dynamical theory of incompressible viscous fluids and the determination of the criterion. *Philosophical Transactions of the Royal Society, 186A*, 123–164.

Reynolds, W. C. (1974). Recent advances in the computation of turbulent flows. *Advances in Chemical Engineering, 9*, 193–246. Prepared for an American Institute of Chemical Engineers short course on turbulence given in November, 1970.

Reynolds, W. C. (1976). Computation of turbulent flows. *Annual Review of Fluid Mechanics, 8*, 183–208.

Richardson, L. F. (1922). *Weather prediction by numerical processes*. Cambridge: Cambridge University Press.

Richardson, L. F. (1926). Atmospheric diffusion shown on distance-neighbour graphs. *Proceedings of the Royal Society, A110*, 709–737.

Riegels, F. (1943). Russische Laminarprofile. Deutsche Luftfahrtforschung, Untersuchungen und Mitteilungen 3040. *ZWB*.

Rogallo, R. S., & Moin, P. (1984). Numerical simulation of turbulent flows. *Annual Review of Fluid Mechanics, 16*, 99–137.

Roland, A. (1985). *Model Research. The National Advisory Committee for Aeronautics 1915–1958*. 2 vols. Washington, DC: NASA. SP-4103.

Roshko, A. (1976). Structure of turbulent shear flows: A new look. *AIAA Journal, 14*(10), 1349–1357.

Roshko, A. (1991). The Mixing Transition in Free Shear Flows. *The Global Geometry of Turbulence. Impact of Nonlinear Dynamics. Edited by Javier Jimenez. Proceedings of a NATO Advanced Research Workshop on The Global Geometry of Turbulence, held July 8-14, 1990, in Rota, Spain*, pp. 3–12.

Rott, N. (1990). Note on the history of the Reynolds Number. *Annual Reviews of Fluid Mechanics, 22*, 1–11.

Rotta, J. C. (1990). *Die Aerodynamische Versuchsanstalt in Göttingen, ein Werk Ludwig Prandtls. Ihre Geschichte von den Anfängen bis 1925*. Göttingen: Vandenhoeck und Ruprecht.

Rouse, H., & Ince, S. (1957). *History of hydraulics*. Iowa Institute of Hydraulic Research: State University of Iowa, Iowa City.

Ruelle, D. (1972). Strange attractors as a mathematical explanation of turbulence. In M. Rosenblatt & C. van Atta (eds.) *Statistical Models and Turbulence. Proceedings of a Symposium held at the University of California, San Diego (La Jolla) July 15–21, 1971*. Springer: Berlin, Heidelberg, 12:292–299. Lecture Notes in Physics.

Ruelle, D. (1980). Strange attractors. *The Mathematical Intelligencer, 2*(3), 126–137.

Ruelle, D. (1991). *Chance and chaos*. Princeton, NJ: Princeton University Press.

Ruelle, D., & Takens, F. (1971). On the nature of turbulence. *Communications in Mathematical Physics, 20*, 167–192.

Sagaut, P. (2005). Large Eddy Simulation for Incompressible Flows. An Introduction. Springer, 3 edition. Original French edition from, (1998). *Introduction à la simulation des grandes échelles pour les écoulements de fluide incompressible* (p. 1998). Berlin, Heidelberg: Springer.

Saph, A., & Schoder, E. (1903). An experimental study of the resistances to the flow of water in pipes. *Transactions of the American Society of Civil Engineering, 51*, 253–312.

Schiller, L. (1921). Experimentelle Untersuchungen zum Turbulenzproblem. *Zeitschrift für Angewandte Mathematik und Mechanik (ZAMM), 1*, 436–444.

Schiller, L. (1925). Das Turbulenzproblem und verwandte Fragen. *Physikalische Zeitschrift, 26*, 566–595.

Schlichting, H. (1941). Der Umschlag laminar/turbulent und das Widerstandsproblem des Tragflügels. Nach einem Vortrag im Seminar der Luftfahrtforschungsanstalt Hermann Göring. Deutsche Luftfahrtforschung, Forschungsbericht Nr. 1410. *ZWB*.

Schlichting, H. (1942). *Vortragsreihe 'Grenzschichttheorie' gehalten im Wintersemester 1941/42 in der Luftfahrtforschungsanstalt Hermann Göring*. ZWB: Braunschweig.

Schlichting, H. (1949a). Lecture Series "Boundary Layer Theory", Part I: Laminar Flows. *NACA TM No. 1217*. Translation of "Vortragsreihe" W.S. 1941/42, Luftfahrtforschungsanstalt Hermann Göring, Braunschweig.

Schlichting, H. (1949b). Lecture Series "Boundary Layer Theory", Part II: Turbulent Flows. *NACA, TM No. 1218*. Translation of "Vortragsreihe" W.S. 1941/42, Luftfahrtforschungsanstalt Hermann Göring, Braunschweig.

Schlichting, H. (1951). *Grenzschicht-Theorie*. Karlsruhe: Braun. völlige Neubearbeitung der 1942 in Typoskriptform erschienenen schriftlichen Ausarbeitung einer 1941/1942 an der Luftfahrt-forschungsanstalt in Braunschweig gehaltenen Vortragsreihe des Verfassers.

Schmaltz, F. (2005). *Kampfstoff-Forschung im Nationalsozialismus: zur Kooperation von Kaiser-Wilhelm-Instituten*. Wallstein, Göttingen: Militär und Industrie.

Schmitt, F. (2017). Turbulence from 1870 to 1920: The Birth of a Noun and of a Concept. *Comptes Rendues Mecanique, 345*, 620–626.

Schubauer, G. B., & Skramstad, H. K. (1943). *Laminar-Boundary-Layer Oscillations and Transition on a Flat Plate* (p. 909). NACA Report No. 909: Originally issued in April 1943 as NACA Advance Confidential Report.

Schultz-Grunow, F. (1940). *Neues Reibungswiderstandsgesetz für glatte Platten. Luftfahrt-forschung, 17*(8), 239–246.

Schumann, U. (1973). *Ein Verfahren zur direkten numerischen Simulation turbulenter Strömungen in Platten- und Ringspaltkanälen und über seine Anwendung zur Untersuchung von Turbulenz modellen*. Ph.D. thesis, TH Karlsruhe.

Schumann, U. (1975). Subgrid scale model for finite difference simulations of turbulent flows in plane channels and annuli. *Journal of Computational Physics, 18*, 376–404.

Schumann, U., & Friedrich, R. (ed.) (1985). *Direct and Large Eddy Simulation of Turbulence*. Springer. Proceedings of the EUROMECH Colloquium No. 199, Munich, FRG, September 30 to October 2, 1985. Wiesbaden: Vieweg+Teubner.

Simmons, L., & Salter, C. (1934). Experimental investigation and analysis of the velocity variations in turbulent flow. *Proceedings of the Royal Society London, A, 145*, 212–234.

Smagorinsky, J. (1963). General circulation experiments with the primitive equations. I: The Basic Experiment. *Monthly Weather Review, 91*(3), 99–164.

Smagorinsky, J. (1983). The beginnings of numerical weather prediction and general circulation modeling: Early recollections. *Advances in Geophysics, 25*, 3–37.

Sommerfeld, A. (1900). Neuere Untersuchungen zur Hydraulik. *Verhandlungen der Gesellschaft Deutscher Naturforscher und Ärzte, 72*, 56.

Sommerfeld, A. (1903). Naturwissenschaftliche Ergebnisse der neueren technischen Mechanik. *Verhandlungen der Gesellschaft deutscher Naturforscher und Ärzte, 75*, 199–216.

Sommerfeld, A. (1909). Ein Beitrag zur hydrodynamischen Erklärung der turbulenten Flüssigkeits-bewegung. Atti del IV Congresso Internazionale dei Matematici (Roma, 6–11 Aprile 1908). Roma. *Tip. della R. Accademia dei Lincei, 3*, 116–124.

Sommerfeld, A. (1945). *Vorlesungen über theoretische Physik, Band II: Mechanik der deformier-baren Medien*. Leipzig: Akademische Verlagsgesellschaft.

Spalding, D. B. (1991). Kolmogorov's two-equation model of turbulence. *Proceedings: Mathematical and Physical Sciences, 434*, 211–216.

Sreenivasan, K. R. (1991). Fractals and multifractal in fluid turbulence. *Annual Review of Fluid Mechanics, 23*, 539–600.

Sreenivasan, K. R. (2011). G. I. Taylor: the inspiration behind the Cambridge school. In P. A. Davidson, Y. Kaneda, K. Moffatt, & K. R. Sreenivasan (Eds.), *A Voyage Through Turbulence* (pp. 127–186). Cambridge: Cambridge University Press.

Sreenivasan, K. R., & Meneveau, C. (1986). The fractal facets of turbulence. *Journal of Fluid Mechanics, 173*, 357–386.

Steen, P., & Brutsaert, W. (2017). Saph and Schoder and the friction law of Blasius. *Annual Reviews of Fluid Mechanics, 49*, 575–582.

Stern, N. (1981). John von Neumann: The Principles of Large-Scale Computing Machines. *Annals of the History of Computing*, *3*, 263–273. Reproduction of a speech on 15 May 1946, delivered at a meeting of the Mathematical Computing Advisory Panel of the Office of Research and Inventions of the Navy Department in Washington, D.C.

Stuart, J. T. (1962). Non-linear effects in hydrodynamic stability. In *Proceedings of the Tenth International Congress of Applied Mechanics*. Stressa. 31 August–7 September 1960 (pp. 63–97). Amsterdam: Elsevier.

Swinney, H. L., Fenstermacher, P. R., & Gollub, J. P. (1977). Transition to turbulence in a fluid flow. In H. Haken (Ed.), *Synergetics* (pp. 60–69). New York: Springer.

Swinney, H. L., & Gollub, J. P. (1978). The transition to turbulence. *Physics Today*, *31*(8), 41–49.

Synge, J. L. (1938). Hydrodynamic stability. *Semicentennial Addresses of the American Mathematical Society*, *2*, 227–269.

Synge, J. L. (1939). The stability of plane poiseuille motion. In J.P. Den Hartog & H. Peters (eds) *Proceedings of the Fifth International Congress on Applied Mechanics* (pp. 326–332). Cambridge MA, New York: John Wiley.

Taylor, G. (1939). Some recent developments in the study of turbulence. In J.P. Den Hartog & H. Peters (eds.) *Proceedings of the Fifth International Congress on Applied Mechanics* (pp. 294–310). Cambridge MA, New York: John Wiley.

Taylor, G. I. (1915). Eddy motion in the atmosphere. *Philosophical Transactions of the Royal Society*, *A215*, 1–26.

Taylor, G. I. (1921). Diffusion by continuous movements. *Peroceedings of the London Mathematical Society*, *20*, 196–212.

Taylor, G. I. (1935a). Statistical theory of turbulence. I. *Proceedings of the Royal Society London*, A, *151*, 421–444.

Taylor, G. I. (1935b). Statistical theory of turbulence. II. *Proceedings of the Royal Society London*, A, *151*, 444–454.

Taylor, G. I. (1935c). Statistical theory of turbulence. III. Distribution of dissipation of energy in a pipe over its cross-section. *Proceedings of the Royal Society London*, A, *151*, 455–464.

Taylor, G. I. (1935d). Statistical theory of turbulence. IV. Diffusion in a turbulent air stream. *Proceedings of the Royal Society London*, A, *151*, 465–478.

Taylor, G. I. (1936). Statistical theory of turbulence. V. Effect of turbulence on boundary layer. Theoretical discussion of relationship between scale of turbulence and critical resistance of spheres. *Proceedings of the Royal Society London*, A, *156*, 307–317.

Taylor, G. I. (1937). The statistical theory of isotropic turbulence. *Journal of the Aeronautical Sciences*, 311–315.

Taylor, G. I. (1946). Sixth international congress for applied mechanics. *Nature*, *4022*, 775.

Tennekes, H., & Lumley, J. L. (1972). *A first course in turbulence*. Cambridge, MA: MIT Press.

Tetervin, N. (1943). Tests in the NACA two-dimensional low-turbulence tunnel of airfoil sections designed to have small pitching moments and high lift-drag ratios. *NACA Confidential Bulletin*, 3I13.

Thomas, L. H. (1952). The stability of plane Poiseuille flow. *Physical Review*, *86*, 812–813.

Thomas, L. H. (1953). The stability of plane Poiseuille flow. *Physical Review*, *91*, 780–783.

Tietjens, O. (1925). Beiträge zur Entstehung der Turbulenz. *Zeitschrift für Angewandte Mathematik und Mechanik (ZAMM)*, *5*, 200–217.

Tobies, R. (1982). Die Gesellschaft für angewandte Mathematik und Mechanik im Gefüge der Wissenschaftsorganisation. *NTM*, *19*, 16–26.

Tobies, R. (2019). *Felix Klein: Visionen für Mathematik, Anwendungen und Unterricht*. Heidelberg: Springer.

Tollmien, W. (1926). Berechnung turbulenter Ausbreitungsvorgänge. *ZAMM*, *6*, 468–478.

Tollmien, W. (1929). Über die Entstehung der Turbulenz. 1. Mitteilung. *Nachrichten von der Gesellschaft der Wissenschaften zu Göttingen, Mathematisch-Physikalische Klasse*, pp. 21–44.

Tollmien, W. (1953). Zusammenfassender Bericht. Fortschritte der Turbulenzforschung. *ZAMM*, *33*(5/6), 200–211.

Toulmin, S. (1972). *Human understanding* (Vol. I). Oxford: Clarendon Press.

Townsend, A. A. (1947). Measurements in the turbulent wake of a cylinder. *Proceedings of the Royal Society, A190*, 551–561.

Townsend, A. A. (1956). *The structure of turbulent shear flow*. Cambridge: Cambridge University Press.

Truesdell, C. (1954). Rational fluid mechanics, 1687–1765. *Editor's introduction to Euleri Opera Omnia, Series II* (vol. 12, pp. IX–CXXV). Zurich: Fuessli.

Tucker, W. A., & Wallace, A. R. (1944). Scale-effect tests in a turbulent tunnel of the NACA 653–418, a=1.0 airfoil section with 0.20-airfoil-chord split flap. *NACA Advance Confidential Report*, L4I22.

Turcotte, D. L. (1988). Fractals in fluid mechanics. *Annual Review of Fluid Mechanics, 20*, 5–16.

Turnbull, D. (1995). Rendering turbulence orderly. *Social Studies of Science, 25*, 9–33.

Vincenti, W. G. (1990). *What engineers know and how they know it: Analytical studies from aeronautical history*. Baltimore: Johns Hopkins University Press.

Vincenti, W. G. (2000). Real-world variation-selection in the evolution of technological form: historical examples. In J. Ziman (Ed.), *Technological Innovation as an Evolutionary Process* (pp. 174–189). Cambridge: Cambridge University Press.

von Karman, T. (1930). Mechanische Ähnlichkeit und Turbulenz. *Nachrichten von der Gesellschaft der Wissenschaften zu Göttingen, Mathematisch-Physikalische Klasse*, pp., 58–76.

von Kármán, T. (1921). Über laminare und turbulente Reibung. *ZAMM, 1*, 233–252.

von Kármán, T. (1930). *Mechanische Ähnlichkeit und Turbulenz* (pp. 58–76). Mathematisch-Physikalische Klasse: Nachrichten von der Gesellschaft der Wissenschaften zu Göttingen.

von Kármán, T. (1931). Mechanische Ähnlichkeit und Turbulenz. *Proceedings of the Third International Congress of Applied Mechanics, 24-29 August 1930*. Edited by A.C.W. Oseen and W. Weibull (3 vol.). AB. Sveriges Litografiska Tryckerier, Stockholm, *1*, 85–93.

von Kármán, T. (1937a). The fundamentals of the statistical theory of turbulence. *Journal of the Aeronautical Sciences, 4*(4), 131–138.

von Kármán, T. (1937b). Turbulence. *Journal of the Royal Aeronautical Society, 41*, 1108–1141. Twenty-fifth Wilbur Wright Memorial Lecture delivered to The Royal Aeronautical Society on May 27th, 1937.

von Kármán, T. (1948). Progress in the statistical theory of turbulence. *Journal of Marine Research, 7*, 252–264.

von Kármán, T. (1967). *The Wind and Beyond*. With Lee Edson. Boston, Toronto: Little, Brown and Company.

von Kármán, T., & Howarth, L. (1938). On the statistical theory of isotropic turbulence. *Proceedings of the Royal Society, 164*, 192–215.

von Mises, R. (1908a). *Theorie der Wasserräder*. Leipzig: Teubner.

von Mises, R. (1908b). Über die Probleme der technischen Hydromechanik. *Jahresbericht der Deutschen Mathematiker-Vereinigung, 17*, 319–325.

von Mises, R. (1912). Kleine Schwingungen und Turbulenz. *Jahresbericht der Deutschen Mathematiker-Vereinigung, 21*, 241–248.

von Mises, R. (1921). Zur Einführung. Über die Aufgaben und Ziele der angewandten Mathematik. *Zeitschrift für Angewandte Mathematik und Mechanik (ZAMM), 1*, 1–15.

von Neumann, J. (1949). Recent theories of turbulence. *John von Neumann Collected Works. Volume VI: Theory of Games, Astrophysics, Hydrodynamics and Meteorology* (pp. 437–472). New York: Pergamon Press.

von Neumann, J., & Richtmyer, R. D. (1950). A method for the numerical calculation of hydrodynamic shocks. *Journal of Applied Physics, 21*(3), 232–237.

Weizsäcker, C. F., & v.,. (1948). Das Spektrum der Turbulenz bei großen Reynoldsschen Zahlen. *Zeitschrift für Physik, 124*, 614–627.

Wieghardt, K. (1944). Zum Reibungswiderstand rauher Platten. Deutsche Luftfahrtforschung, Untersuchungen und Mitteilungen 6612. *ZWB*.

Wieghardt, K. (1948). Wärmeübergang. *Naturforschung und Medizin in Deutschland 1939-1946. Für Deutschland bestimmte Ausgabe der FIAT Review of German Science*, 5(V), 129–133.

Wien, W. (1900). *Lehrbuch der Hydrodynamik*. Leipzig: Hirzel.

Winsberg, E. (2010). *Science in the age of computer simulation*. Chicago: University of Chicago Press.

Yaglom, A. M. (1994). A. N. Kolmogorov as a fluid mechanician and founder of a school in turbulence research. *Annual Reviews of Fluid Mechanics*, 26, 1–22.

Yaglom, A. M. (2012). *Hydrodynamic instability and transition to turbulence*. New York: Springer.

Index

© The Author(s), under exclusive license to Springer Nature Switzerland AG 2019
M. Eckert, *The Turbulence Problem*,
SpringerBriefs in History of Science and Technology,
https://doi.org/10.1007/978-3-030-31863-5

Printed in the United States
By Bookmasters